Luckie

The authors present a new edition of their highly successful introductory textbook. The book has been enlarged and fully revised. Through clear and concise text, attractive presentation and the use of beautiful colour plates, the biology student is drawn into this fascinating introduction to the photosynthetic process.

The student is taken on an entertaining tour of the photosynthetic process. The authors discuss photosynthesis at both macro- and molecular level, placing new ideas in the context of past, present and future research. The role of photosynthesis as a source of food and fuel is highlighted. The student is also encouraged to think 'practically', with a useful chapter on simple laboratory experiments.

The book will appeal to students and teachers of biology, working for advanced school examinations or for college or university degrees.

D0162025

Photosynthesis

The Institute of Biology aims to advance both the science and practice of biology. Besides providing the general editors for this series, the Institute publishes two journals *Biologist* and *the Journal of Biological Education*, conducts examinations, arranges national and local meetings and represents the views of its members to government and other bodies. The emphasis of the *Studies in Biology* will be on subjects covering major parts of first-year undergraduate courses. We will be publishing new editions of the 'bestsellers' as well as publishing additional new titles.

Titles available in this series

An Introduction to Genetic Engineering, D. S. T. Nicholl

Photosynthesis, 5th edition, D. O. Hall and K. K. Rao

Photosynthesis
Fifth edition

D. O. Hall
Professor of Biology
King's College, University of London

and

K. K. Rao
Honorary Lecturer in Biology
King's College, University of London

Published in association with the Institute of Biology

Published by the Press Syndicate of the University of Cambridge
The Pitt Building, Trumpington Street, Cambridge CB2 1RP
40 West 20th Street, New York, NY 10011–4211, USA
10 Stamford Road, Oakleigh, Melbourne 3166, Australia

First edition published by Edward Arnold 1972 (0537 9024)
Second edition 1977
Third edition 1977
Third edition 1981
Fourth edition first published by Edward Arnold 1987 and first
published by Cambridge University Press 1992 (0 521 42806 8)
Fifth edition 1994
Reprinted with revisions 1995

Printed in Great Britain at the University Press, Cambridge

A catalogue record for this book is available from the British Library

Library of Congress cataloguing in publication data

Hall, D. O. (David Oakley)
Photosynthesis / D. O. Hall, K. K. Rao. – 5th ed.
    p.   cm. – (Studies in biology)
'Published in association with the Institute of Biology.'
Includes bibliographical references (p.   ) and index.
ISBN 0 521 43036 4 (hc). – ISBN 0 521 43622 2 (pbk.)
1. Photosynthesis.   I. Rao, K. K.   II. Institute of Biology.
III. Title.   IV. Series.
QK882.H19   1994
581.1'3342 – dc20   93-37397 CIP

ISBN 0 521 43036 4 hardback
ISBN 0 521 43622 2 paperback

SE

Contents

General preface to the series   xi
Preface to the fifth edition   xiii

1 Importance and role of photosynthesis   1
  1.1 Ultimate energy source   1
  1.2 The carbon dioxide cycle   2
  1.3 Efficiency and turnover   4
  1.4 Spectra   5
  1.5 Quantum theory   6
  1.6 Energy units   7
  1.7 Measurement of photosynthetic irradiance   8
  1.8 Some techniques used in photosynthesis research   10

2 History and progress of ideas   22
  2.1 Early discoveries   22
  2.2 Limiting factors   24
  2.3 Light and dark reactions; flashing light experiments   26
  2.4 Further discoveries and formulations   27

3 Photosynthetic apparatus   32
  3.1 Isolation of chloroplasts from leaves   37
  3.2 Chloroplast pigments   39
  3.3 The photosynthetic unit   51
  3.4 Photosynthetic apparatus of C_4 plants   53

4 Light absorption and the two photosystems   57
4.1 Dissipation of absorbed light energy: photochemistry, fluorescence, phosphorescence and thermoluminescence   58
4.2 Energy transfer or sensitized fluorescence   64
4.3 Action spectra, quantum yield   65
4.4 Emerson effect and the two light reactions   67
4.5 Reaction centres and primary electron acceptors   72
4.6 Experimental separation of the two photosystems   72
4.7 Inside-out and right-side-out chloroplast vesicles   74
4.8 Photosynthetic oxygen evolution   76

5 Photosynthetic electron transport and phosphorylation   79
5.1 Reduction and oxidation of electron carriers   80
5.2 Two types of photosynthetic phosphorylation   80
5.3 Non-cyclic electron transport and phosphorylation   81
5.4 ATP synthesis in chloroplasts: the chemiosmotic hypothesis   86
5.5 Cyclic electron transport and phosphorylation   91
5.6 Structure–function relationships   94
5.7 Artificial electron donors, electron acceptors, and inhibitors   95

6 Carbon dioxide fixation: the C_3 and C_4 pathways   99
6.1 Experimental techniques   99
6.2 The photosynthetic carbon reduction (Calvin) cycle   102
6.3 Structure–function relationships   107
6.4 Energetics of CO_2 fixation   107
6.5 Sucrose and starch synthesis   109
6.6 The C_4 (Kortschak, Hatch–Slack) pathway of CO_2 fixation   111
6.7 Crassulacean acid metabolism: CAM species   114
6.8 Light-coupled reactions of chloroplasts other than CO_2 fixation   118
6.9 Photorespiration and glycollate metabolism   121
6.10 Environmental factors affecting CO_2 assimilation by plants   123

7 Bacterial photosynthesis   126
7.1 Classification   126
7.2 Photosynthetic pigments and apparatus   127

7.3 Photochemistry and electron transport   132
7.4 Carbon dioxide fixation   133
7.5 Light energy conversion by halobacteria   135
7.6 Ecological significance of phototrophic bacteria 135
7.7 A comparison of plant and bacterial electron transport 137
7.8 Evolution of photosynthesis   142

8 Research in photosynthesis   145
8.1 Phytochromes   145
8.2 Protoplasts and cells   148
8.3 Origin and development of chloroplasts   148
8.4 Chloroplast genetics; expression and regulation of
     genes; transgenic plants   150
8.5 Transport and assembly of cytoplasmically assembled
     polypeptides into the chloroplast membranes;
     exchange of ions and metabolites through the
     chloroplast envelope   153
8.6 Chloroplast structure   156
8.7 The photosystem II oxygen-evolving reaction   159
8.8 Photosystem II: structure and function   162
8.9 Photosystem I   165
8.10 The cytochrome b_6f complex: the Q cycle   170
8.11 RuBisCO: structure and function   171
8.12 Fluorescence as a probe for energy transfer and stress
     physiology in photosynthesis   175
8.13 Photoinhibition   179
8.14 Energy redistribution between the two photosystems 185
8.15 Role of light in the regulation of photosynthesis: the
     ferredoxin–thioredoxin control system   187
8.16 Whole plant studies and bioproductivity   189
8.17 Photosynthesis and the greenhouse effect   191
8.18 Mimicking photosynthesis   193

9 Laboratory experiments   195
9.1 Reference books for experiments   195
9.2 Photosynthesis in whole plants and algae   195
9.3 Preparation of protoplasts, chloroplasts and
     subchloroplast membranes   196
9.4 Separation and estimation of photosynthetic pigments
     and proteins   197

9.5 Measurement of photosynthetic electron transport
      using oxygen electrode and/or spectrophotometer   197
9.6 Proton flux and photophosphorylation   198

Appendix I   Chemical names   199
Appendix II   Abbreviations and prefixes used in the text 200
Further reading   202
   Non-specialist books   202
   More specialized books   202
   Reviews and articles   203
   More specialized articles   207

Index   208

Plate section is between pp. 66 and 67

General preface to the series

Charged by its Royal Charter to promote biology and its understanding, the Institute of Biology recognises that it is not possible for any one text book to cover the entirety of a course. If evidence was needed, the success of the *Studies in Biology* series was a testimony to the need for specialist, up-to-date publications in education. The Institute is therefore pleased to collaborate with Cambridge University Press in producing a new title in the *Studies in Biology* series.

The new series is set to provide as great a boon to the new generation of students as the original did to their parents.

Suggestions and comments from readers will always be welcomed and should be addressed either to the Studies in Biology Editorial Board at Cambridge University Press or c/o The Books Committee at the Institute.

Robert Priestley
The General Secretary

The Institute of Biology
20–22 Queensberry Place
London SW7 2DZ

Preface to the fifth edition

The first edition of this book appeared in 1972. Photosynthesis research has expanded rapidly in the last two decades; some earlier concepts have been modified or even abandoned. 'Recent discoveries' mentioned in the first edition have now become common knowledge. Side by side with the understanding of the biophysics and molecular biology of photosynthetic processes, research is also now directed towards the application of photosynthesis in solving some of the problems facing humanity. Design of transgenic and hybrid plants by genetic engineering, clean (solar) energy technology, production and use of biomass as fuel, and alleviation of global warming by CO_2 sequestration with plants and algae are some of the current areas in photosynthesis research. The interest in photosynthesis is reflected by the fact that over a thousand researchers now participate in the triennial International Photosynthesis Congresses.

In this fifth edition we have retained the style and features of the previous editions which seem to appeal to students and teachers alike, namely an introduction to the basic ideas of photosynthesis, an historical outline as to how these ideas developed, the current status in our understanding of photosynthesis, and an overview as to where future research will be oriented. At the same time we have taken this opportunity to revise and update many of the sections and to expand the book by the inclusion of new topics and figures. A new section on techniques used in photosynthetic research has been introduced with a brief description and illustrations of the equipment employed. The chapter on bacterial photosynthesis has been expanded, by popular demand, with additional figures and a detailed section comparing bacterial and plant-type photosynthesis.

The chapter describing research in photosynthesis is expanded to include recent advances in chloroplast genetics, photosystem I and II structure, oxygen evolution, photoinhibition, and fluorescence. New sections on phytochromes, cyt b f complex, RuBisCO structure, and photosynthesis and the greenhouse effect are added.

The chapter suggesting laboratory experiments has been revised with addition of new experiments and reference books. The reference list has also been updated.

A novel feature of the Fifth edition is the introduction of colour plates illustrating the structures of cyanobacteria, chloroplasts, chloroplast membrane components, X-ray structure of RuBisCO, electron transport through the photosynthetic bacterial reaction centre, and of some instruments used for measuring photosynthesis in the field. We are grateful to ADC Ltd; Hansatech Instruments, and LI-COR, Inc. for sponsorship.

Once again we are indebted to colleagues who very kindly and promptly provided us with colour prints, electron micrographs and figures. The suggestions from readers and reviewers of the previous editions have been helpful in the preparation of this edition. We again look forward to further comments on the contents of this new edition.

D. O. Hall
K. K. Rao
London

1

Importance and role of photosynthesis

1.1 Ultimate energy source

Hardly a day goes by without the importance of photosynthesis being brought to our attention. All our food and our fossil and biological fuels (biomass) are derived from the process of photosynthesis, both past and present. Increasingly, the products of photosynthesis are being sought to feed and fuel the world and also to provide chemicals and fibres. How our changing environment will affect plants and their productivity has become a topic of great interest. Thus an understanding of the fundamental and applied aspects of photosynthesis is now essential to a wide range of scientists and technologists – from agriculture and forestry through ecology and biology to chemistry, genetics and engineering. It is this universality that attracts varied approaches to studying photosynthesis and makes it such an exciting field of work to so many different types of people. We hope that this becomes evident in our book.

The term photosynthesis literally means building up or assembly by light. As used commonly, photosynthesis describes the process by which plants synthesize organic compounds from inorganic raw materials in the presence of sunlight. All forms of life in this universe require energy for growth and maintenance. Algae, higher plants and certain types of bacteria capture this energy directly from the solar radiation and utilize the energy for the synthesis of essential food materials. Animals cannot use sunlight directly as a source of energy; they obtain the energy by eating plants or by eating other animals which have eaten plants. Thus the ultimate source of all metabolic energy in our planet is the sun, and photosynthesis is essential for maintaining all forms of life on earth.

We use coal, natural gas, petroleum, etc. as fuels. All these fuels are decomposition products of land and marine plants or animals, and the energy stored in these materials was captured from the solar radiation millions of years ago. Solar radiation is also responsible for the formation of wind and rain, and hence the energy from windmills and hydro-electric power stations could also be traced back to the sun.

The major chemical pathway in photosynthesis is the conversion of carbon dioxide and water to carbohydrates and oxygen. The reaction can be represented by the equation:

$$CO_2 + H_2O \xrightarrow[\text{plants}]{\text{sunlight}} \underset{\text{carbohydrate}}{[CH_2O]} + O_2$$

The carbohydrates formed possess more energy than the starting materials, namely CO_2 and H_2O. By the input of the sun's energy, the energy-poor compounds, CO_2 and H_2O, are converted to the energy-rich compounds, carbohydrates and O_2. The energy levels of the various reactions which lead up to the overall equation above can be expressed on an oxidation–reduction scale ('redox potential' given in volts) which tells us the energy available in any given reaction – this will be discussed later in Chapter 4. Photosynthesis can thus be regarded as a process of converting radiant energy of the sun into chemical energy of plant tissues.

1.2   The carbon dioxide cycle

The CO_2 content of the atmosphere remained almost constant for millenia in spite of its depletion during photosynthesis. However, there has been a 27% increase since the Industrial Revolution in the last century, resulting in the so-called greenhouse effect. All plants and animals carry out the process of respiration (in mitochondria) whereby O_2 is taken from the atmosphere by living tissue to convert carbohydrates and other tissue constituents eventually to CO_2 and water, with the simultaneous liberation of energy. The energy is stored in ATP (adenosine triphosphate) and is utilized for the normal functions of the organism. Respiration thus causes a decrease in the organic matter and O_2 content and an increase in the CO_2 content of the planet. Respiration by living organisms and combustion of carbonaceous fuels consume on average about 10 000 tonnes of O_2 every second on the surface of the earth. At this rate, all the O_2 of the atmosphere would have been used up in about 3000 years. Fortunately for us, the loss of organic

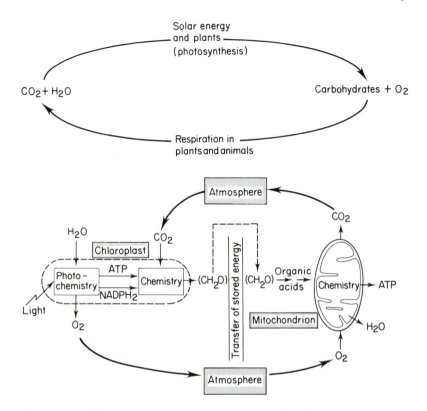

Fig. 1.1   The CO_2 and O_2 cycle in the atmosphere and the cell.

matter and atmospheric oxygen during respiration is counterbalanced by the production of carbohydrates and oxygen during photosynthesis. Under ideal conditions, the rate of photosynthesis in the green parts of plants is about 30 times as much as the rate of respiration in the same tissues. Thus photosynthesis is very important in regulating the O_2 and CO_2 content of the earth's atmosphere. The cycle of operations can be represented as shown in Fig. 1.1. All the CO_2 in the atmosphere is cycled through plants, via photosynthesis, every 300 years, and all the O_2 is cycled every 2000 years.

It should be made clear that the energy liberated during respiration is finally dissipated from the living organism as heat and is not available for recycling. Thus for millions of years, energy has been constantly removed from the sun and wasted as heat in the earth's atmosphere. But fortunately there is still enough energy available from the sun for photosynthesis to continue for many hundreds of millions of years.

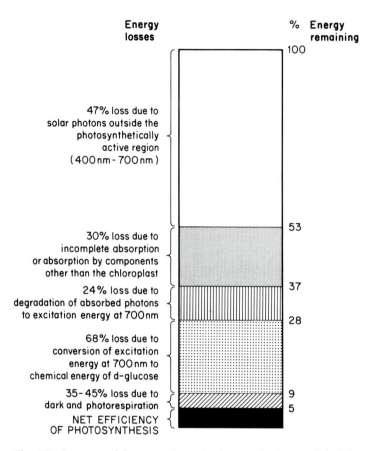

Fig. 1.2   Summary of the energy losses in photosynthesis as sunlight falls on a leaf at 25°C.

1.3   Efficiency and turnover

Photosynthetic efficiency on a global basis may be defined as the fraction of photosynthetically active radiation (PAR) that falls on the earth's surface which is converted to *stored* energy by photosynthesis in the biosphere.

The solar energy striking the earth's atmosphere every year is equivalent to about 56×10^{23} joules (J) of heat. Of this, roughly half is reflected back by the clouds and by the gases in the upper atmosphere. Of the remaining radiation that reaches the earth's surface, only 50% is in the spectral region of light that could bring about photosynthesis, the other half being weak infra-red radiation (Fig. 1.2). Thus the annual influx of energy of photosynthetically active radiation, i.e. from violet to red light, to the earth's surface is equivalent to about 15×10^{23} J. However, some 40% of this is reflected by ocean surface, deserts, etc. and only the rest can be absorbed by the plant life on land and sea. Recent estimates of the total annual amount of biomass (plant matter produced by photosynthesis) are about 2×10^{11} tonnes of organic matter, which is equivalent to about 4×10^{21} J of energy. Thus the average coefficient of utilization of the incident photosynthetically active radiation by the entire flora of the earth is only about 0.27% ($4 \times 10^{21}/15 \times 10^{23}$). The annual food intake by the human population (approximately 5300 million at present) is about 990 million tonnes, or 16×10^{18} J, which is only 0.4% ($16 \times 10^{18}/4 \times 10^{21}$) of the annual biomass produced. It is interesting that the total consumption of energy by the world, including biomass, in 1991 was about 3.9×10^{20} J – this was only a tenth of the energy stored by photosynthesis! In fact, the energy content of the biomass standing on the earth's surface today (80% trees) is equivalent to about two-thirds of our proven reserves of fossil fuel (36×10^{21} J), i.e. oil, gas and coal. The total resources of fossil fuel stored under the earth's surface (260×10^{21} J) represent only 60 years of net photosynthesis.

1.4   Spectra

Light is a form of electromagnetic radiation. All electromagnetic radiation has wave characteristics and travels at the same speed of 3×10^8 m s^{-1} (c, the speed of light). But the radiations differ in wavelength, the distance between two successive peaks of the wave. Gamma rays and X-rays have very small wavelengths (less than one thousand millionth of a centimetre, 10^{-11} m), while radio waves are in the order of 10^4 cm. Wavelengths of visible light are conveniently expressed by a unit called a nanometre. One nanometre is one thousand millionth of a metre (1 nm $= 10^{-9}$ m). It has been known since the time of Isaac Newton that white light can be separated into a spectrum, resembling the rainbow, by passing light through a prism. The visible portion of this spectrum ranges from the violet at about 380 nm to the far red at 750 nm (Fig. 1.3).

The atmosphere of the sun consists mainly of hydrogen. The energy of the sun is derived from the fusion of four hydrogen nuclei to form a helium nucleus. The fusion process is a multistep reaction which can be simplified as $4H \rightarrow He + h\nu$ (energy). The mass of the He nucleus is less than the total mass of 4H; the mass lost during fusion is converted to energy and emitted as photons or quanta. The energy liberated during the nuclear fusion

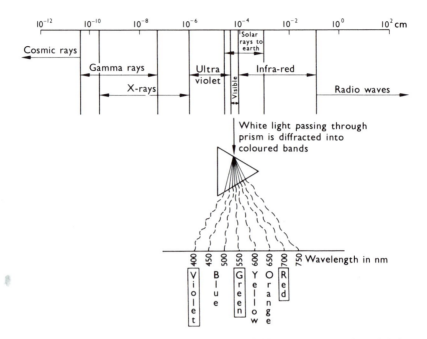

Fig. 1.3   Spectra of electromagnetic radiation and diffraction pattern of visible light.

maintains the surface temperature of the sun around 6000 K. The sun radiates energy representing the entire electromagnetic spectrum, but the earth's atmosphere is transparent only to part of the infra-red and ultraviolet light and all the visible light. The ultraviolet waves, which are somewhat shorter than the shortest visible light waves, are absorbed by the oxygen and ozone of the upper atmosphere. This is fortunate since ultraviolet radiations are harmful to living organisms. At 6000 K, the temperature of the sun, the maximum intensity of emitted light lies in the orange part of the visible spectrum, around 600 nm.

1.5   Quantum theory

In 1900 Max Planck enunciated the theory that the transfer of radiation energy within a hot object occurred in discrete 'units' of energy called quanta. Planck's quantum theory can be expressed mathematically as $E = h\nu$ where E is the energy of a single quantum of radiation, ν is the frequency of the radiation (frequency is the number of waves transmitted per unit time), and h a constant. The Planck's constant (h) has the dimensions of the product of energy and time and its value in the c.g.s. system is 6.626×10^{-34} J s. Planck's theory proposes that an oscillator of fundamental frequency (v) would take up energy hv, $2hv$, $3hv \rightarrow nhv$, but it could not acquire less than a whole number of energy quanta. Five years later, Albert Einstein extended Planck's theory to light and proposed that light energy is transmitted not in a continuous stream but only in individual units or quanta. The energy of a single quantum of light, or *photon* (luxon before 1916), is the product of the frequency of light and Planck's constant, i.e. $E = hv$. Since frequency is inversely related to wavelength, it follows that photons of short-wave light are more energetic than photons of light of longer wavelength, i.e. photons of blue light (400 nm) at one end of the visible spectrum are more energetic than those of red light (700 nm) at the other end.

For photosynthesis to take place the pigments present in photosynthesizers should absorb the energy of a photon at a characteristic wavelength and then utilize this energy to initiate a chain of photochemical and chemical events. We will learn later that an electron is ejected from the reaction centre pigment immediately after the absorption of a suitable quantum of light. It should be emphasized that a photon cannot transfer its energy to two or more electrons, nor can the energy of two or more photons combine to eject an electron. Thus the photon should possess a critical energy to excite a single electron from the pigment molecule and initiate photosynthesis. This accounts for the low efficiency of infra-red radiation in plant photosynthesis since there is insufficient energy in the quantum of infra-red light. Certain bacteria, however, contain pigments which absorb infra-red radiation and carry out a type of photosynthesis which is quite different from plant-type photosynthesis in that no O_2 is evolved during the process (see Chapter 7).

1.6   Energy units

According to Einstein's law of photochemical equivalence, a single molecule will react only after it has absorbed the energy of one photon (hv). Hence 1 mole (gram-molecule) of a compound must absorb the energy of N photons (N $= 6.023 \times 10^{23}$, the Avogadro number), i.e. Nhv, to start a reaction. The total energy of photons absorbed by 1 mole of a compound is called an Einstein, i.e. one Einstein $= 6.023 \times 10^{23}$ quanta.

Let us calculate the energy of a mole (or Einstein) of red light of wavelength 650 nm (6.5×10^{-7} m). The frequency, $v = c/\lambda =$ speed of light/ wavelength of light. The speed of light is 3×10^8 m s^{-1}.

Table 1.1. *Energy levels of visible light*

Wavelength (nm)	Colour	Energy level J/mole	kcal/mole	eV/photon
700	Red	17.10×10^4	40.87	1.77
650	Orange-red	18.40×10^4	43.98	1.91
600	Yellow	19.95×10^4	47.68	2.07
500	Blue	23.95×10^4	57.24	2.48
400	Violet	29.93×10^4	71.53	3.10

$$\nu = 3.0 \times 10^8/6.5 \times 10^{-7} = 4.61 \times 10^{14}$$

$$E = N h \nu,$$

i.e. energy = number of molecules × Planck's constant × frequency

$$\therefore\ E = 6.023 \times 10^{23} \times 6.626 \times 10^{-34} \times 4.61 \times 10^{14} = 18.40 \times 10^4 \text{ joules}$$
= energy of one Einstein (or mol) of red light
or $E = 18.40 \times 10^4/4.184 \times 10^3 = 43.98$ kcal

(One kilocalorie, kcal, is equal to 4.184×10^3 joules.) Thus 1 mole of red light at 650 nm contains 18.40×10^4 joules of energy.

The energy of photons can also be expressed in terms of electron volts. An electron volt, eV, is the energy acquired by an electron when it falls through a potential of 1 volt, which is equal to 1.6×10^{-19} joules. If 1 molecule of a substance acquires an average energy of 1 eV, the total energy of a mole can be calculated to be 9.64×10^4 joules ($1.6 \times 10^{-19} \times 6.023 \times 10^{23}$). Thus the energy of 1 mole of 650 nm light (Table 1.1) is equal to 1.91 eV ($18.40 \times 10^4/9.64 \times 10^4$).

1.7   Measurement of photosynthetic irradiance

Historically, light intensity was measured in terms of *lumens* (lm), a lumen being defined as the luminous flux on a unit surface, all points of which are at unit distance from a uniform point source of one candle. Intensity of illumination (illuminance) was expressed either as foot candles (lm ft^{-2}) or *lux* (lm m^{-2}).

Nowadays, photobiologists prefer to measure light energy incident on a surface, i.e. radiant flux density or *irradiance*, in terms of the units of power as watts per sq metre (W m^{-2}). Since photochemical reactions in photosyn-

Table 1.2. *Terminology for radiation data*

Term	Unit	Definition
Radiant energy	J	Energy in the form of electromagnetic radiation
Radiant flux	$W = J s^{-1}$	Radiant energy emitted or absorbed by a surface per unit time
Radiant flux density	$W m^{-2}$	Radiant flux (of a specific wavelength region) incident on a small sphere divided by the cross-sectional area of the sphere
Irradiance	$W m^{-2}$	Radiant flux incident on a unit area of plane surface per unit time
Fluence	$mol\ m^{-2}$	Number of photons incident across a unity area of plane surface
Photon flux density (PFD)	$mol\ m^{-2} s^{-1}$	Photon flux per unit area
Photosynthetically active radiation (PAR)		Solar radiation in the 400 to 700 nm waveband
Photosynthetic photon flux density (PPFD)	$mol\ m^{-2} s^{-1}$	Flux of 400 to 700 nm solar radiation

thesis depend more on the *number* of photons incident on a surface rather than on the energy content of these photons, it is more logical to express photosynthetic irradiance in terms of the number of quanta (photons) falling on unit surface in unit time, i.e. as the *photon flux density*. The photon (or quantum) flux density (Q) in a particular wavelength region is measured in units of $mol\ m^{-2} s^{-1}$ where a mol is 6.023×10^{23} (Avogadro's number) quanta or photons. Since an Einstein (E) is defined as 6.023×10^{23} quanta, Q can also be expressed as $E\ m^{-2} s^{-1}$. A more practical unit to express photosynthetic photon flux density (PPFD) is $\mu mol\ m^{-2} s^{-1}$ or $\mu E\ m^{-2} s^{-1}$. For example, the solar irradiance reaching the earth's surface in full sunlight is approximately $1000\ W m^{-2}$ or 100 000 lux, in which the PPFD is $2000\ \mu mol\ m^{-2} s^{-1}$. There are no factors to convert radiation data directly between photometric, radiometric and quantum units. Approximate conversion factors are given on pp. 442 and 443 of Hall *et al.* (1993). Some of the terms used in plant physiology to express radiation are summarized in Table 1.2.

There are different types of instruments to measure irradiance – a typical one used to measure PPFD in quanta is shown in Fig. 1.4.

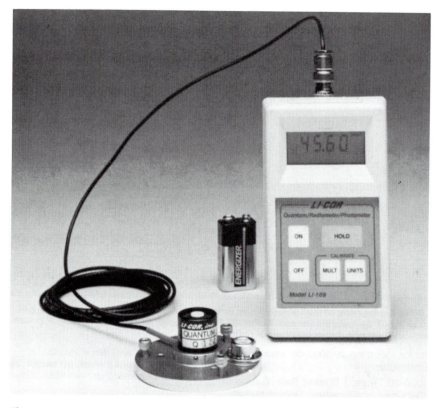

Fig. 1.4   A portable quantum radiometer/photometer for measurement of photon flux in the field or in the laboratory. (Courtesy: LI-COR, Inc., Lincoln, Nebraska, USA.)

1.8   Some techniques used in photosynthesis research

The oxygen electrode   The oxygen electrode (Fig. 1.5) is a convenient instrument to measure uptake or liberation of O_2 during a reaction. The electrode works on the principle of polarography and is sensitive enough to detect O_2 concentrations of the order of 10^{-8} moles cm^{-3} (0.01 millimolar). The apparatus consists of a platinum wire sealed in plastic as cathode, and an anode of circular silver wire bathed in a saturated KCl solution. The electrodes are separated from the reaction mixture by an O_2 gas-permeable teflon membrane. The reaction mixture in the plastic (or glass) container is stirred constantly with a small magnetic stirring rod. When a voltage is applied across the two electrodes, with the platinum electrode negative to

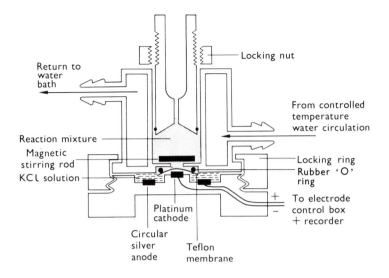

Fig. 1.5   The oxygen electrode. (Rank Bros., Bottisham, Cambridge.)

the reference electrode, the oxygen in the solution undergoes electrolytic reduction. The flow of current in the system between 0.5 and 0.8 V varies in a linear relationship to the partial pressure of the oxygen in solution. The instrument is usually operated at a voltage of about 0.6 V. The current liberated is measured by connecting the electrode to a suitable recorder. The whole apparatus is kept at a constant temperature by circulating water from a controlled temperature water source. The effects of light and of various chemicals on photosynthesis are measured using the oxygen electrode.

The Clark-type electrode shown in Fig. 1.5 is generally used to measure O_2-exchange activities of isolated chloroplasts and of algal and cyanobacterial cells in the laboratory. Modified versions of the electrode which can be used to measure O_2 evolution from leaves in the field are available.

Infra-red gas analyzer (IRGA): measurement of CO_2 exchange   The IRGA is a reliable, sensitive and convenient instrument to determine photosynthetic CO_2 assimilation and photorespiratory CO_2 release in plants. Heteroatomic gas molecules (CO_2, H_2O, NH_3 etc.) absorb infra-red (IR) radiation at specific wavelengths, whereas gas molecules containing identical atoms (O_2, N_2, Ar_2) do not absorb IR radiation. The main absorption peak for CO_2 is at 4.26 μm, with minor peaks at 2.66, 2.77 and 14.99 μm. Since water vapour (a usual constituent of air) also absorbs at 2.77 μm, CO_2

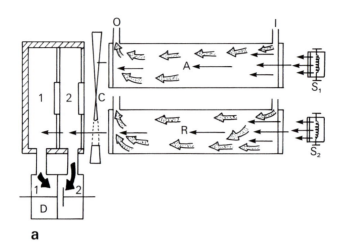

a

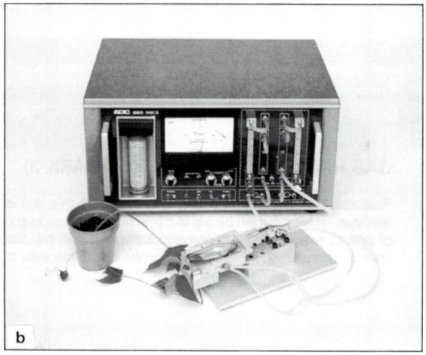

b

Fig. 1.6 (**a**) Schematic illustration of a double-beam IRGA with absorption cells 1 and 2 and the detector D in series. Infra-red radiation from the sources S_1 and S_2 is chopped (at C) alternately so that the radiation transmitted initially from the analysis cell (A) and then the reference cell (R) is received in turn by both the detector cells. There is a continuous flow of the sample and the reference gases through the cells A and R respectively; both cells are provided with inlet (I) and outlet (O) for gas flow. (See also Long and Hällgren in Hall *et al.*, 1993.) (**b**) The ADC 225 Mark 3 Gas Analyser for continuous measurements of CO_2 or H_2O in the laboratory using the IRGA technique. (Courtesy: The Analytical Development Co. Ltd, Hoddesdon, UK.)

measurements are usually made in an IRGA by the use of interference filters which preferentially isolate the 4.3-μm absorption band for detection.

An IRGA consists of three basic parts (Fig. 1.6a): 1. a low-voltage IR beam source which is usually a spiral of Nichrome alloy or tungsten heated to about 800°C; 2. a gas analysis cell with a gas inlet and outlet through which the IR beam is passed; and 3. a solid state IR radiation detector. An internal pump in the instrument passes the gas sample and CO_2-free air (reference standard), alternately, through the cell for periods of 2 seconds each. The IR radiation leaving the cell is passed through a filter and the signal arising in one half-cycle is stored and compared with the signal received in the next half-cycle. The difference in the signal size reaching the detector between the half-cycles (of pumping sample and reference gases) is proportional to the amount of CO_2 in the gas sample. The CO_2 assimilation rate is expressed as the amount of CO_2 assimilated per unit leaf area (or other photosynthesizing tissue) and time: e.g. μmol CO_2 consumed m^{-2} s^{-1}.

Most IRGAs are calibrated before gas analysis using CO_2-free air and a sample containing a precisely known concentration of CO_2 in the range found in the analysis sample. Portable commercial versions of IRGAs are shown in Fig. 1.6b and Plate VI.

Fluorescence spectrophotometer: chlorophyll fluorescence   Fluorescence is an analytically useful emission spectroscopic technique in which atoms or molecules are excited by the absorption of electromagnetic radiation; the excited species then relax to the ground state, dissipating their excess energy as photons (light). The fluorescence absorption and emission processes occur within nano (10^{-9}) seconds. During the photosynthetic processes of light absorption and energy transduction, a small percentage of the absorbed light, not utilized in photochemistry, is re-emitted as fluorescence (for more details refer to §4.1 and 8.12). Chlorophyll fluorescence data have been dubbed as the 'plant physiologist's stethoscope' because they are extremely useful in the diagnosis of photosynthetic activity of plants under normal and 'stressed' conditions. A block diagram of an advanced version of a continuous excitation fluorescence spectrophotometer which can be used in the laboratory or in the field is given in Fig. 1.7 (see also Plate VII).

Electron paramagnetic resonance (EPR) spectroscopy   Electron paramagnetic resonance (EPR), also known as electron spin resonance (ESR), is a magnetic technique used for the study of unpaired electron spins that are

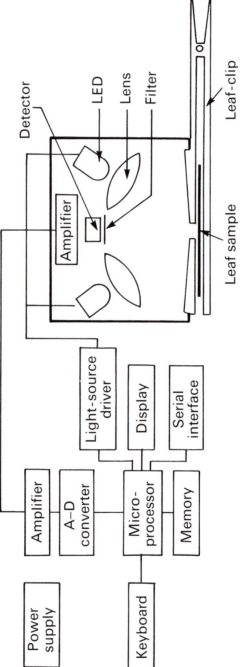

Fig. 1.7 Measurement of photosynthesis in the field by fluorescence technique using the Hansatech portable Plant Efficiency Analyser (PEA). LED, light-emitting diode. (Courtesy: Hansatech Instruments Ltd; King's Lynn, UK.)

found in organic free radicals and in paramagnetic species such as the transition metal ions (Cu^{2+}, Fe^{3+}, Mn^{2+} etc.). The EPR spectrum of a compound can provide information about its mobility, oxidation state of the metal ions, spin state, nature of the ligands, co-ordination geometry of the paramagnetic centre, etc.

In a typical EPR spectrometer, the sample is placed in a microwave cavity and scanned in a magnetic field. Resonant absorption of microwave, of frequency v, by the free electrons in the sample takes place and is detected with a diode (Fig. 1.8). The condition for resonant absorption is given by the equation: $hv = g\beta H_0$ where h is the Planck's constant, v the frequency, β the Bohr magneton (electron magnetic moment), H_0 the applied magnetic field, and g is a characteristic factor for the sample under examination. The equation can be rewritten as $g = \dfrac{hv}{\beta H_0}$. The g value in an EPR spectrum is a spectroscopic variable which is a characteristic of the paramagnetic component and can be used as a fingerprint to identify the component.

EPR spectra are conveniently displayed as a first derivative of the microwave absorption spectrum against applied magnetic field.

One of the earliest applications of EPR in biology was the detection of two light-induced EPR signals (named signals 1 and 2) in spinach chloroplasts, by Commoner in 1956 in the USA, which were later identified as originating from photosystems I and II (PSI and PSII), respectively. As will be discussed in later chapters, EPR has been successfully applied in photosynthesis for: the detection and identification of bacterial and chloroplast photosystem components; the valence state of Mn and the composition of S states in the water oxidation complex; chloroplast and carotenoid triplet states, etc.

Electron microscopy: ultrastructure of cells and organelles   Cells of photo-synthetic prokaryotes and chloroplasts, phycobilisomes, chromatophores and other photosynthetic complexes can only be seen under a microscope. Light microscopes have a resolving power of about 200 nm which is high enough to view prokaryotic cells and isolated organelles such as chloro-plasts and chlorosomes. In order to observe the ultrastructure and composition of photosynthetic membranes, electron microscopes are used.

In an electron microscope (EM), electron beams are focused on a specimen by electromagnets and the scattered electron patterns (which are characteristic of the specimen under observation) are recorded in a signal detector (Fig. 1.9). Because electrons have a much shorter wavelength (pico (10^{-12}) metres) compared to visible light (nano (10^{-9}) metres), the

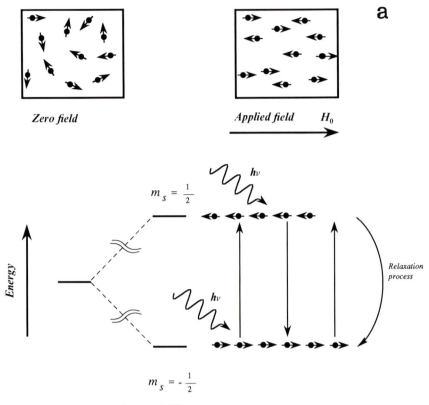

Zero field

Applied field H_0

$m_s = \frac{1}{2}$

hv

Energy

hv

Relaxation process

$m_s = -\frac{1}{2}$

The resonance condition $hv = g\beta H_0$

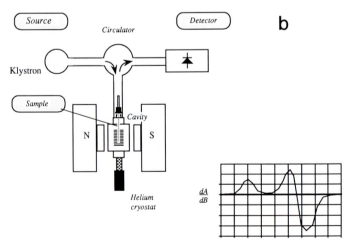

Source

Circulator

Detector

b

Klystron

Sample

Cavity

N   S

Helium cryostat

$\frac{dA}{dB}$

Magnetic field

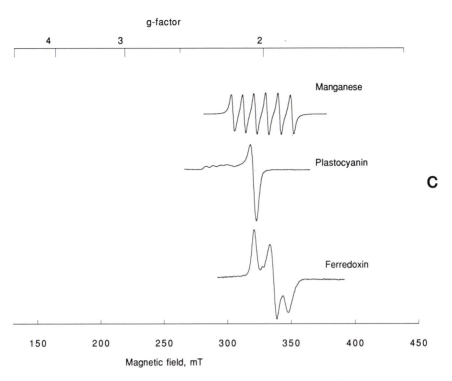

g-factor

4   3   2

Magnetic field, mT

150   200   250   300   350   400   450

Fig. 1.8   Diagram illustrating (**a**) the principles of EPR spectroscopy, (**b**) the operation of a low temperature EPR spectrometer, and (**c**) typical EPR spectra of paramagnetic species. Note that the Mn EPR Spectrum is not from the PSII water-oxidizing complex. (Courtesy: R. Cammack, King's College, London.)

resolving power of the EM is more than 100 times that of the light microscope, and magnifications of up to 200 000 times can be obtained. There are two basic types of electron microscopes: the *scanning electron microscope* (SEM) and the *transmission electron microscope* (TEM).

The SEM is useful for the detailed study of the external surface features of micro-organisms and organelles. The ultrastructure of chloroplast and thylakoid membrane surfaces has been observed by SEM. The sample is first chemically fixed by glutaraldehyde (2–4% buffered solution), followed, if necessary, by post-fixation in 1% osmium tetroxide. After dehydration in an ethanol/water series, the samples are mounted onto stubs, critical point dried and coated with carbon, gold or palladium. Cells of prokaryotes can also be observed without fixation; the sample is prepared by critical point drying only. The stub is then placed in the EM chamber, the chamber evacuated, and an electron beam focused into a fine

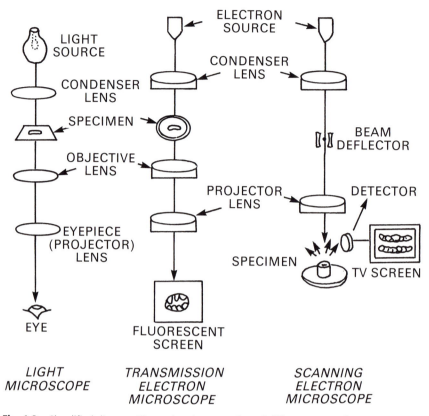

Fig. 1.9   Simplified diagram illustrating the operation of different types of microscopes.

probe and swiftly passed back and forth over the specimen. Complete scanning of the surface takes only a few seconds. In the SEM image, holes and fissures in the specimen will appear dark, and ridges and knobs appear light, creating a three-dimensional effect. The SEM can resolve details up to 10 nm size and produce 10 000 fold magnification.

The TEM can resolve features in biological specimens up to 2 nm size and has a much higher magnification capacity than the SEM. However, the protocol for specimen preparation for TEM is more elaborate and tedious. In the TEM, electrons are actually passed through the specimen, and because electrons have very limited penetrating power, the plastic block holding the specimen must be very thin (50–100 nm thick). Such thin sections are practically two-dimensional slices of material and the image may fail to convey a three-dimensional perspective. For TEM examination, the sample is chemically fixed, dehydrated and then entrapped in plastic. Thin slices of the plastic are cut using a microtome and then mounted on copper grids, dried and examined.

Freeze-etching   Freeze-etching or freeze-fracturing is a technique for preparation of EM samples in which frozen cells or biological membranes are broken open under high vacuum and the newly exposed surfaces are replicated on thin (2–10 nm) carbon or metal film. Freeze-fracture EM of cells of phototrophic bacteria and of chloroplast membranes enables us to view the ultrastructure on both surfaces and within the plane of the membrane. Generally for freeze-etching, the biological specimen is fixed by physical means, i.e. by *rapid* cooling to $-150°C$ to $-190°C$.

A typical protocol for freeze-fracture EM is as follows. The sample (e.g. isolated chloroplast membranes), suspended in a buffered medium containing 2% glycerol (which prevents formation of ice crystals during freezing), is quickly frozen by plunging into liquid Freon or propane cooled in liquid N_2 and then stored in liquid N_2. The solid block is transferred to the cold 'stage' of a freezing microtome installed in a high vacuum chamber. After evacuation, the block, which is kept at $-100°C$, is sectioned, the specimen dried by sublimation (water removal) and a thin layer of Pt-C deposited over the specimen to produce a replica. The replica is cleaned by washing with potassium hypochlorite solution (bleach) and examined in the TEM. The resulting electron micrograph, with an attainable resolution of <1 nm, provides a three-dimensional image equivalent to the living state. Freeze-fracture EM of PSII-enriched chloroplast membranes has shown that the three extrinsic polypeptides associated with the O_2-evolving apparatus are located on the lumenal surface of the thylakoid membrane.

Ligands and immunospecific antibodies can be prepared with metal labels such as gold or ferritin and then incorporated into surfaces of cells or isolated organelles (chlorosomes, chloroplasts etc.) prior to freeze-etching. Electron micrographs of immuno-gold-labelled thylakoid membranes have provided evidence for the distribution of the cytochrome *bf* complex in three domains of the photosynthetic membrane.

Applications of microscopes in the structure determination in photosynthesis are illustrated in Fig. 1.10.

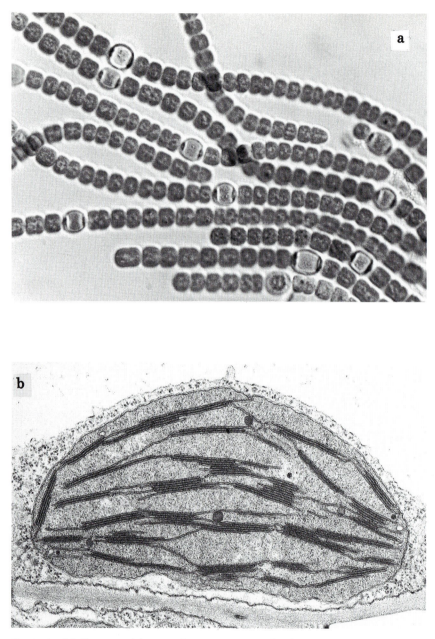

Fig. 1.10 (**a**) Filaments of the cyanobacterium *Nostoc flagelliforme* as observed under a light microscope. The larger cells are the heterocysts and the others vegetative. Magnification 950 x. (Courtesy: R. Lichtl, King's College, London.) (**b**) Tobacco chloroplasts as revealed by scanning electron microscopy; the grana stacks can be distinguished from the stromal lamellae. Magnification 18 960 x. (Courtesy: L.A. Staehelin, University of Colorado, Boulder, Colorado.) (**c**) Transmission electron

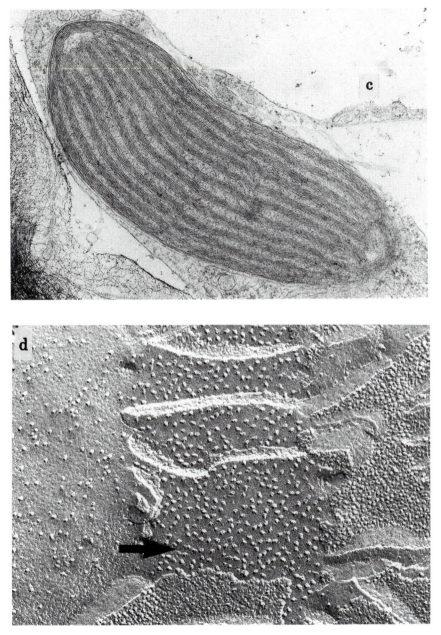

micrograph of chloroplasts from the brown alga *Laminaria abyssalis*. Magnification 39 425 x. (Courtesy: D. Strbac and J. F. Pacy, King's College, London.) (**d**) The surface of pea thylakoid as revealed by freeze-etch electron microscopy; the stacked membrane regions in the centre of the micrograph can be differentiated from unstacked area by the multimeric particles (arrow) that protrude from the background. Magnification 85 540 x. (Courtesy: L.A. Staehelin.)

2

History and progress of ideas

2.1 Early discoveries

In the early half of the seventeenth century the Flemish physician van Helmont grew a willow tree in a bucket of soil, feeding the soil with rain water only. He observed that after 5 years the tree had grown to a considerable size, though the amount of soil in the bucket had not diminished significantly. Van Helmont naturally concluded that the material of the tree came from the *water* used to wet the soil. In 1727 the English botanist Stephen Hales published a book *Vegetable Staticks*, in which he observed that plants used mainly *air* as the nutrient during their growth. Between 1771 and 1777 the English chemist Joseph Priestley (who was one of the discoverers of oxygen) conducted a series of experiments on combustion and respiration and came to the conclusion that green plants were able to reverse the respiratory processes of animals. Priestley burnt a candle in an enclosed volume of air and showed that the resultant air could no longer support burning. A mouse kept in the residual air died. A green branch of mint, however, continued to live in the residual air for weeks. At the end of this time Priestley found that a candle could burn in the reactivated air and a mouse could breathe in it. We now know that the burning candle used up the *oxygen* of the enclosed air, which was replenished by the photosynthesis of the green mint. A few years later (in 1779) the Dutch physician, Jan Ingenhousz, discovered that plants evolved oxygen *only in sunlight* and also that only the *green* parts of the plant carried out this process.

In 1782, Senebier, a Swiss minister, confirmed the findings of Ingenhousz and observed further that plants used as nourishment *carbon dioxide* 'dissolved in water'. Early in the nineteenth century another Swiss scholar, de Saussure, studied the quantitative relationships between the CO_2 taken up by a plant and the amount of organic matter and O_2 produced, and came to the conclusion that *water* was also consumed by plants during assimilation of CO_2. In 1817 two French chemists, Pelletier and Caventou, isolated the green substance in leaves and named it *chlorophyll*. Another milestone in the history of photosynthesis was the enunciation in 1845 by Mayer, a German physician, that plants transform energy of sunlight into chemical *energy*. By the middle of the last century the phenomenon of photosynthesis could be represented by the relationship:

$$CO_2 + H_2O + light \xrightarrow[\text{plant}]{\text{green}} O_2 + \text{organic matter} + \text{chemical energy}$$

Accurate determinations of the ratio of CO_2 consumed to O_2 evolved during photosynthesis were carried out by the French plant physiologist Boussingault. He found in 1864 that the photosynthetic ratio – the volume of O_2 evolved to the volume of CO_2 used up – is almost unity. In the same year, the German botanist Sachs (who also discovered plant respiration) demonstrated the formation of *starch* grains during photosynthesis. Sachs kept some green leaves in the dark for some hours to deplete them of their starch content. He then exposed one half of a starch-depleted leaf to light and left the other half in the dark. After some time the whole leaf was exposed to iodine vapour. The illuminated portion of the leaf turned dark violet due to the formation of starch–iodine complex; the other half did not show any colour change.

The direct connection between oxygen evolution and chloroplasts of green leaves, and also the correspondence between the action spectrum of photosynthesis and the absorption spectrum of chlorophyll (see Chapter 4), were demonstrated by the German botanist Engelmann in the 1880s. He placed a filament of the green alga *Spirogyra*, with its spirally arranged chloroplasts, on a microscope slide together with a suspension of oxygen-requiring, motile bacteria (Fig. 2.1). The slide was kept in a closed chamber in the absence of air and illuminated. Motile bacteria would move towards regions of greater O_2 concentration. After a period of illumination the slide was examined under a microscope and the bacterial population counted. Engelmann found that the bacteria were concentrated around the green bands of the algal filament. In another series of experiments he illuminated the alga with a spectrum of light by interposing a prism between the light source and the microscope stage. The largest number of bacteria sur-

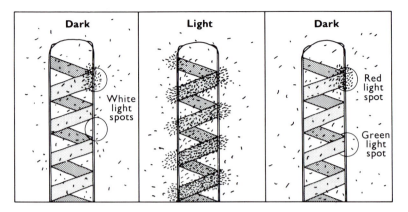

Fig. 2.1   Summary of Engelmann's experiment for studying photosynthesis using the alga *Spirogyra* and motile bacteria. The alga has a spiral chloroplast and the bacteria migrate towards regions of higher O_2 concentration. **Left.** Illumination with a spot of white light. **Centre.** Complete illumination with white light. **Right.** Illumination with spots of red and green light. Note the lack of O_2 evolution in green light.

rounded those parts of the algal filament that were in the blue and red regions of the spectrum. The chlorophylls present in the alga absorbed blue and red light; since light has to be absorbed to bring about photosynthesis, Engelmann concluded that chlorophylls are the active photoreceptive *pigments* for photosynthesis. The state of knowledge of photosynthesis at the beginning of this century could be represented by the equation:

$$(CO_2)_n + H_2O + light \xrightarrow[\text{plant}]{\text{green}} (O_2)_n + starch + chemical\ energy$$

It was not clear whether the source of O_2 was water or CO_2.

2.2   Limiting factors (see also §6.10 and 8.16)

Although by the beginning of this century the overall reaction of photosynthesis was known, the discipline of biochemistry had not advanced enough to understand the mechanism of reduction of CO_2 to carbohydrates. It should be admitted that even now we know very little about certain aspects of photosynthesis. Early attempts were made to study the effects of light intensity, temperature, CO_2 concentration, etc. on the overall yields of photosynthesis. Though plants of divergent species were used in these

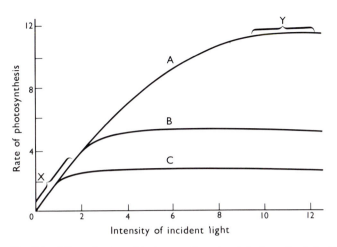

Fig. 2.2   Effect of external factors on the rate of photosynthesis in *Chlorella*. **A**, Effect of light intensity at 25°C and 0.4% CO_2; **B**, at 15°C and 0.4% CO_2; **C**, at 25°C and 0.01% CO_2. All units in the graph are arbitrary.

studies, most of the determinations were carried out with unicellular green algae, *Chlorella* and *Scenedesmus*, and the unicellular flagellate *Euglena*. Unicellular algae are more suitable for quantitative research since they can be grown in all laboratories under fairly standard conditions throughout the year. They can be suspended uniformly in aqueous buffer solutions, and aliquots of the suspension can be transferred with a pipette as though they were true solutions.

The extent of photosynthesis performed by a plant depends on a number of internal and external factors. The chief internal factors are the structure of the leaf and its chlorophyll content, the accumulation within the chloroplasts of the products of photosynthesis, the influence of enzymes, and the presence of minute amounts of mineral constituents. The external factors are the quality and quantity of light incident on the leaves or algae, the ambient temperature, and the concentration of CO_2 and O_2 in the surrounding atmosphere and, more generally, the availability of water and nutrients.

The effect of light intensity

The data obtained from early studies on the effect of external factors on the photosynthetic activity of a healthy suspension of *Chlorella* cells are illustrated in Fig. 2.2. At low light intensities the rate of photosynthesis, as measured by O_2 evolution, increases linearly in proportion to light intensity. This region of the curve, marked X, is known as the light-limiting region. With more and more light intensity, photosynthesis under near-atmospheric conditions (21% O_2 and 0.034% CO_2) becomes less efficient until, after about 10 000 lux (about 100 W m^{-2}), increasing light intensity produces no further effect on the rate of photosynthesis. This is indicated by the horizontal parts of the curves in the figure. This plateau region, designated Y, is the light-saturation region. If the rate of photosynthesis is to be raised in this region, factors other than light intensity would have to be adjusted.

The amount of sunlight falling on a clear summer day in many places on the earth is about 1000 W m^{-2}. Thus except for plants growing in thick forests and in shade, there is often sufficient sunlight incident on the plants to saturate their photosynthetic capacity, if other factors are not limiting.

Effect of temperature

A comparison of the curves A and B in the figure shows that at low light intensities the rate of photosynthesis is the same at 15°C and at 25°C. The reactions in the light-limiting region, like true photochemical reactions, are not sensitive to temperature. At higher light intensities, however, the rate of photosynthesis is much higher at 25°C than at 15°C. Thus, factors other than mere photon absorption influence photosynthesis in the light-saturation region.

Effect of CO_2 concentration

In the light-limiting region, the rate of photosynthesis is not affected by lowering the CO_2 concentration, as shown by curve C in Fig. 2.2. Thus, it can be inferred that CO_2 does not participate directly in the photochemical reaction. But at light intensities above the light-limiting region, photosynthesis is appreciably enhanced by increasing the CO_2 concentration.

2.3   Light and dark reactions; flashing light experiments

As early as 1905 the British plant physiologist Blackman interpreted the shape of the light-saturation curves by suggesting that photosynthesis is a two-step mechanism involving a photochemical or '*light*' reaction and a non-photochemical or '*dark*' reaction. The dark reaction, which is enzymatic, is slower than the light reaction and hence at high light intensities the rate of photosynthesis is entirely dependent upon the rate of the dark reaction. The light reaction has a low or zero temperature coefficient, while the dark reaction has a high temperature coefficient, characteristic of enzymatic reactions. It should be clearly understood that the so-called dark reaction can proceed both in light and in dark, with multiple steps.

The light and dark reactions can be separated by using flash illuminations lasting fractions of a second. Light flashes lasting less than a millisecond (10^{-3} s) can be produced either mechanically by placing a slit in a rotating disc in the path of a steady light beam, or electrically by loading up a condenser and discharging it through a vacuum tube. Ruby lasers emitting red light at 694 nm are also used as a radiation source. In 1932 Emerson and Arnold, in Illinois, illuminated suspensions of *Chlorella* cells with condenser flashes lasting about 10^{-5} s. They measured the rate of oxygen evolution in relation to the energy of the flashes, the duration of the dark intervals between the flashes, and the temperature of the cell suspension. Flash saturation occurred in normal cells when one molecule of O_2 evolved from 2500 chlorophyll molecules. Emerson and Arnold concluded that the maximum yield of photosynthesis is not determined by the number of chlorophyll molecules capturing the light but by the number of enzyme molecules which carry out the dark reaction. They also observed that for dark time intervals (between successive flashes) greater than 0.06 s, the yield of O_2 per flash was independent of the dark time interval; the yield per light flash increased with dark time intervals from 0 to 0.06 s. Thus the dark reaction determining the saturation rate of photosynthesis takes about 0.06 s for completion. The average dark reaction time was calculated to be about 0.02 s at 25°C.

2.4   Further discoveries and formulations

Prior to about 1930, many investigators in the field believed that the primary reaction in photosynthesis was splitting of CO_2 by light to carbon and O_2; the carbon was subsequently reduced to carbohydrates by water in a different series of reactions. Two important discoveries in the 1930s changed this viewpoint. Firstly, a variety of bacterial cells were found to assimilate CO_2 and synthesize carbohydrates without the use of light energy. Then the Dutch microbiologist van Niel, who worked mostly in

California, in comparative studies of plant and bacterial photosynthesis, showed that some bacteria can assimilate CO_2 in light without evolving O_2. Such bacteria would not grow photosynthetically unless they were supplied with a suitable hydrogen donor substrate. Photosynthesis could be represented, according to van Niel, by the general equation:

$$CO_2 + 2H_2A \xrightarrow[\text{chlorophyll}]{\text{light}} (CH_2O) + H_2O + 2A$$

where H_2A is the oxidizable substrate. Van Niel suggested that photosynthesis of green plants and algae is a special case in which H_2A is H_2O and $2A$ is O_2. The primary photochemical act in plant photosynthesis would be the splitting of water to yield an oxidant (OH) and a reductant (H). The primary reductant (H) could then bring about the reduction of CO_2 to cell materials, and the primary oxidant (OH) could be eliminated through a reaction to liberate O_2 and reform H_2O. The overall equation of photosynthesis for green plants, after van Niel, is:

$$CO_2 + 4H_2O \xrightarrow[\text{chlorophyll}]{\text{light}} (CH_2O) + 3H_2O + O_2$$

which is a sum of three individual steps:

(*i*) $4H_2O \xrightarrow[\text{green pigments}]{\text{light}} 4(OH) + 4H$

(*ii*) $4H + CO_2 \rightarrow (CH_2O) + H_2O$

(*iii*) $4(OH) \rightarrow 2H_2O + O_2$

The reaction sequences clearly show that the oxygen is evolved from water and not from CO_2.

The Hill reaction   The second important observation was made in 1937 by R. Hill of Cambridge University. Hill separated the photosynthesizing particles (chloroplasts) of green leaves from the respiratory particles by differential centrifugation of a homogenate of leaf tissues. Hill's chloroplasts did not evolve O_2 when illuminated as such (due to possible damage of the chloroplasts during isolation) but did so when suitable electron acceptors (oxidants) like potassium ferrioxalate or potassium ferricyanide were added to the illuminated suspension. One molecule of O_2 was evolved for every four equivalents of oxidant reduced photochemically. Later, many quinones and dyes were found to be reduced by illuminated chloroplasts. This phenomenon, now known as the Hill reaction, is a light-driven transfer of electrons from water to non-physiological oxidants (Hill reagents) against the chemical potential gradient. The significance of the Hill reaction lies in the demonstration of the fact that photochemical O_2 evolution can be separated from CO_2 reduction in photosynthesis.

The decomposition of water, and the resulting liberation of O_2 during photosynthesis, was established by Ruben and Kamen in California in 1941. They exposed photosynthesizing cells to water enriched in the recently available oxygen isotope of mass 18 (^{18}O). The isotopic composition of the O_2 evolved was the same as that of the water and not that of CO_2 used. Kamen and Ruben also discovered the radioactive isotope ^{14}C, which was successfully used by later workers to study photosynthetic CO_2 assimilation.

The O_2-evolving chloroplasts prepared by Hill failed to photoreduce CO_2, the natural electron acceptor of photosynthesis. It was then thought that CO_2 assimilation is impossible in a cell-free system – a view held by many until the early 1950s. A series of (43 consecutive!) experiments conducted in Arnon's laboratory in California disproved this hypothesis. The photosynthetic reduction of NADP to $NADPH_2$ by isolated chloroplasts with simultaneous evolution of O_2 was reported in 1951 by three different laboratories, including Arnon's. Then in 1954, Arnon, Allen and Whatley demonstrated photophosphorylation (i.e. light-induced synthesis of ATP from ADP and Pi), and also the simultaneous assimilation of $^{14}CO_2$ and evolution of O_2 by isolated spinach chloroplasts; cell-free photosynthesis was thus accomplished! (See Chapter 5.) (Frenkel, in 1954, had observed the formation of ATP from ADP and Pi on illumination of membrane preparations from photosynthetic bacteria.) Arnon and co-workers (including Trebst, Losada and Tsujimoto) soon distinguished the 'light' and 'dark' reactions of chloroplasts by separating the chloroplasts into granal and stromal fractions and demonstrating ATP and $NADPH_2$ formation in the grana in light, and the dark enzymatic CO_2 reduction by enzymes of the stroma. Later the group discovered cyclic photophosphorylation which produced only ATP, and proposed its role for providing additional ATP for CO_2 fixation in chloroplasts.

Meanwhile Bassham, Benson and Calvin in California were studying the incorporation of $^{14}CO_2$ into photosynthesizing algae (see Chapter 6). They showed that the reduction of CO_2 to sugars proceeded by dark enzymic reactions and also that two molecules of $NADPH_2$ and three molecules of ATP were required for the reduction of every molecule of CO_2. For

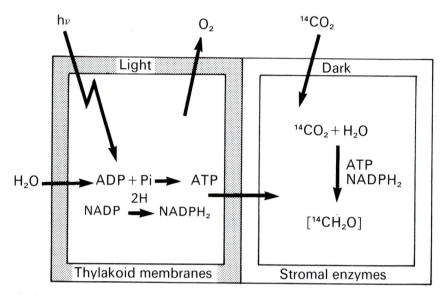

Fig. 2.3 Oxygenic photosynthesis as visualized in the 1960s.

elucidating the path of photosynthetic carbon reduction, Calvin was awarded the Nobel Prize in 1961.

Emerson and co-workers in Illinois measured quantum yields of photosynthesis in algae and observed that the average quantum yield obtained by using two superimposed light beams of different wavelengths was higher than the average quantum yield obtained by using the two beams separately (§4.3). To explain this enhancement of quantum yield, Emerson and Rabinowitch in 1960 postulated the existence of two light reactions in photosynthesis. In the same year, Hill and Bendall put forward the Z scheme of photosynthesis showing the operation of two photosystems in series in photosynthetic electron transport and phosphorylation. In 1961, Mitchell in Britain put forward the chemiosmotic theory to explain ATP formation in biological membranes (see Chapter 5). The state of knowledge in photosynthesis at the beginning of the 1960s can be summarized as shown in Fig. 2.3.

In 1965, Kortschak *et al.* in Hawaii, studying the incorporation of $^{14}CO_2$ into sugar cane leaves, reported that in sugar cane the first stable compounds formed in photosynthesis are malic and aspartic acids, and these acids are subsequently converted to sucrose via 3-phosphoglyceric acid and hexose phosphates. These findings were extended and confirmed by Hatch and Slack in Australia, who observed (in 1966) that after exposure of sugar cane leaf segments to $^{14}CO_2$ for 1 s, more than 93% of the fixed radioactivity was located in the C_4 dicarboxylic acids malic, aspartic and oxaloacetic. Thus the C_4 pathway of CO_2 fixation was discovered, which is also referred to as the Kortschak, Hatch–Slack pathway.

The most notable achievement in photosynthesis in recent years is the crystallization of the reaction centre from a purple photosynthetic bacterium and the determination of its structure by X-ray crystallographic analysis by Deisenhofer, Huber and Michel in Germany, for which they were awarded the Nobel Prize in 1988.

Crystallization of the light harvesting complex and photosynthetic reaction centres from selected plants and cyanobacteria have been achieved. High resolution X-ray analysis of these crystals in the 'dark' and 'light' phases would help to unravel the structure–function relationships of the reaction centre components and to understand the molecular processes associated with O_2 evolution.

3

Photosynthetic apparatus

The photosynthetic apparatus is that part of the leaf or algal cell which contains the ingredients for absorbing light and for channelling the energy of the excited pigment molecules into a series of chemical and enzymatic reactions. Engelmann's experiments (Chapter 2) have shown that chlorophylls are the pigments responsible for capturing light quanta. Knowledge concerning the subcellular structure in which chlorophyll is located comes from light and electron microscopy, and from cell fractionation techniques. In green algae and in higher plants the chlorophyll is contained in a cellular plastid called the *chloroplast*.

In leaves, chloroplasts occupy about 8% of the total cell volume. In exceptional cases, such as in grass, more than 30 million chloroplasts per cm^2 of surface area can be found.

Electron microscope pictures show that chloroplasts in higher plants, e.g. spinach, tobacco, are saucer-shaped bodies 4 to 10 μm in diameter and 1 μm in thickness (1 μm $= 10^{-6}$ m) with an outer membrane or envelope separating them from the rest of the cytoplasm (Fig. 3.1). The number of chloroplasts per cell, in higher plants, varies from 1 to more than 100, depending upon the particular plant and on the growth conditions. In many plants the chloroplasts are able to reproduce themselves by a simple division.

Internally the chloroplast is comprised of a system of *lamellae* or flattened *thylakoids* which are arranged in stacks in dense green regions known as *grana* (Fig. 3.2). Each lamella in the chloroplast may contain two double-layer membranes 5–7 nm thick. The grana are embedded in a colourless matrix called the *stroma* and the whole chloroplast is surrounded by a

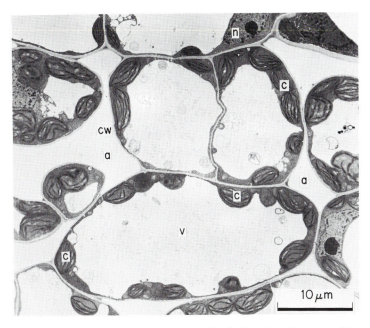

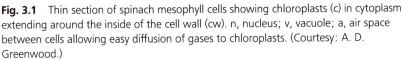

Fig. 3.1 Thin section of spinach mesophyll cells showing chloroplasts (c) in cytoplasm extending around the inside of the cell wall (cw). n, nucleus; v, vacuole; a, air space between cells allowing easy diffusion of gases to chloroplasts. (Courtesy: A. D. Greenwood.)

bounding double membrane, the *chloroplast envelope*. Within the chloroplast the grana are interconnected by a system of loosely arranged membranes called the stroma lamellae. The detailed structure of the thylakoids is shown in Fig. 3.3. These models are based on electron microscopy using freeze-fracturing techniques. The surfaces thus exposed show the distribution of chlorophyll–protein complexes embedded in or associated with the lipid bilayer, which forms the 'backbone' of the membrane. The organization of these complexes, which are seen as particles, is different in the stacked grana membrane regions from that in the unstacked stroma membranes. This is explained by predominance of the photosystem II oxygen-evolving reaction complexes in the grana and the photosystem I particles in the stroma lamellae (see Fig. 8.4).

The space between the thylakoid double membrane, the intrathylakoid region, is termed the *lumen*. The lumenal space is about 5–10 nm wide; its full contents are not known. During photosynthetic electron transport, protons from the stroma are transferred to the lumenal space through the thylakoid membrane; these protons are used in ATP synthesis.

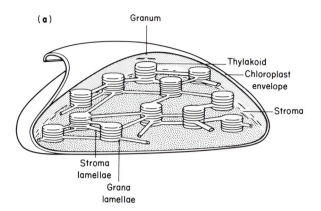

(a)

Granum

Thylakoid
Chloroplast
envelope

Stroma

Stroma
lamellae

Grana
lamellae

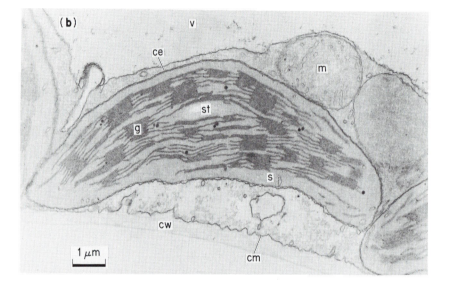

(b)

v

ce

m

st

g

s

cw

cm

1 μm

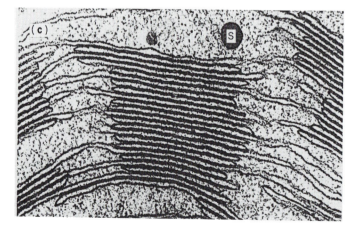

(c)

S

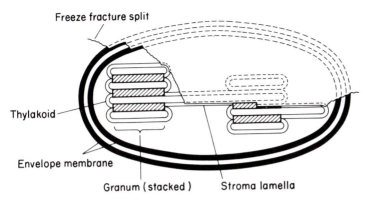

Freeze fracture split

Thylakoid

Envelope membrane

Granum (stacked)   Stroma lamella

Fig. 3.3 Diagram of chloroplast structure as revealed by freeze-etching in the electron microscope. Freeze-fracturing of the chloroplast results in cross-fracturing or splitting of the lamellae, thus exposing interior regions of the membrane. (See also Fig. 8.4.)

It is possible to fractionate the chloroplasts of higher plants so that the green lamellae are separated from the colourless stroma matrix. The lamellar membranes in which the chlorophyll is embedded are approximately half lipid and half protein in chemical composition. The proteins catalyze the enzyme reactions and give mechanical strength to the membranes. Most of the light-harvesting chlorophylls a and b (Chl a and Chl b) are conjugated with specific membrane proteins. The presence of lipids in the membranes facilitates energy storage and offers selective permeability of sugars, salts, substrates, etc. Chloroplast lipids play an important role in maintaining membrane structure and function. One of the causes for the thermal and photodecay of chloroplasts is the release of lipids from the membranes and their oxidation.

The quantum conversion of light energy and associated electron transport reactions of photosynthesis occur in the lamellae. The stroma contains many soluble proteins, including the enzymes of the Calvin cycle (see Chapter 6) which carry out the dark phase reduction of CO_2 to carbohydrates.

Fig. 3.2 (*left*) (**a**) Cut-away representation of a chloroplast to show three-dimensional structure. (**b**) Section of a chloroplast in the cytoplasm of a spinach leaf cell. ce, chloroplast envelope; g, granum consisting of stacks of thylakoids; s, stroma; st, starch granule in chloroplast; cm, cytoplasmic membrane; cw, cell wall; m, mitochondrion; v, vacuole, (**c**) A single granum within a chloroplast showing stacks of thylakoids and interconnecting stroma lamellae between granal stacks. s, lipid droplet in the stroma. (Courtesy: A. D. Greenwood.)

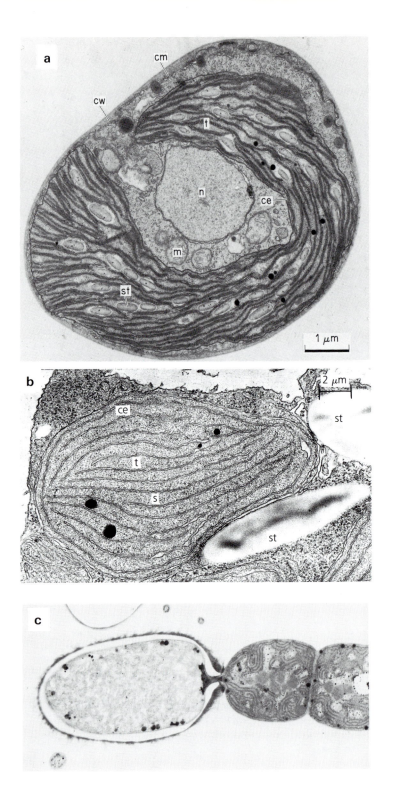

The lamellar structure is found not only in chloroplasts of higher plants but also in algal chloroplasts. In algae the shapes of the chloroplast are varied. The most primitive algae, the blue-greens (now called cyanobacteria since they are prokaryotic organisms) do not contain chloroplasts as such. The photosynthetic material present in these organisms consists of parallel layers of lamellar membranes traversing the cytoplasm. Figure 3.4 shows chloroplasts from a green and a red alga and the cell structure of a blue-green alga. The structure of chloroplast isolated from a fern is shown in Plate III.

3.1   Isolation of chloroplasts from leaves

The first photosynthetically active chloroplasts were isolated by Hill; these preparations were active only in O_2 evolution coupled to the reduction of non-physiological electron acceptors. Arnon and Whatley isolated chloroplasts in isotonic sodium chloride, i.e. about 0.35 M or 2%. Their preparations were capable of photoreduction of NADP and photophosphorylation but were able to fix CO_2 only at low rates, although they contained all the enzymes of the Calvin CO_2-fixation cycle. These chloroplasts appeared intact under the light microscope but electron microscope pictures of the preparation indicated that they had lost their outer membranes and were naked lamellar systems. Such preparations are called 'Type C' (broken) chloroplasts (Fig. 3.5). Walker in Sheffield has developed techniques for the isolation of 'Type A' (complete) chloroplasts which retain their outer envelopes and which can fix CO_2 at rates up to 90% of those of whole leaves. (See Hall (1972) *Nature*, **235**, 125, for discussion of chloroplast Types.)

Two methods used in our laboratory for the preparation of chloroplasts are given below. All solutions and apparatus used should be pre-cooled in ice. The preparation should be carried out as quickly as possible.

Fig. 3.4   (a) The green alga *Coccomyxa* sp., which is a symbiont within a lichen, showing a single cup-shaped chloroplast within the cell. The thylakoids (t) are in groups of three. ce, chloroplast envelope; st, starch granule within chloroplast; cw, cell wall; cm, cytoplasmic membrane; n, nucleus; m, mitochondrion. (Courtesy: H. Bronwen Griffiths.) (b) Chloroplast of the red alga *Ceramium* sp. with single thylakoids (t) lying nearly parallel to one another in the stroma (s). ce, chloroplast envelope; st, starch granule outside chloroplast. Dark spots in chloroplast are lipid droplets. (Courtesy: A. D. Greenwood.) (c) Vegetative and heterocyst cells of the cyanobacterium *Anabaena azollae*, a symbiont in the *Azolla* fern from rice fields. X 5000 (Courtesy: D. J. Shi.)

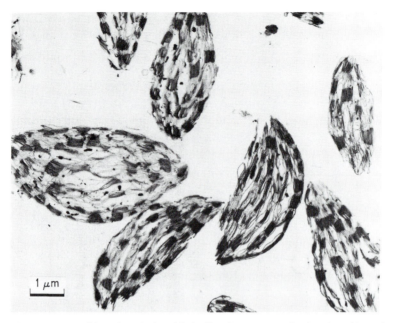

Fig. 3.5 Bean chloroplasts isolated in buffered sucrose media showing chloroplasts with intact thylakoids but without a chloroplast envelope – Type C as defined in the text. (Courtesy: A. D. Greenwood and R. Leech, University of York.)

Preparation 1   Procedure of Whatley and Arnon (modified). Grinding medium: 0.35 M NaCl; 0.04 M tris/HCl buffer pH 8.0.

Cut 25 g of spinach (*Spinacea oleraceae*) leaves into small pieces 0.5–1 cm long. Place in a laboratory or domestic blender with 50 cm^3 grinding medium. Blend for 10 s on low and 20 s on high speed. Filter the homogenate through a nylon bag (or four layers of cheesecloth) into a centrifuge tube. Centrifuge 4 min at 2000 g. Discard the supernatant. Re-suspend the pellet in 2 cm^3 0.35 M NaCl using a small piece of absorbent cotton wool wrapped round the end of a glass rod. This preparation consists of Type C (broken) chloroplasts. Chloroplast fragments (Type E) are prepared by diluting the suspension with 10 vol. of water to give a NaCl concentration of 0.035 M.

Preparation 2   Procedure of Walker.
Grinding medium:
Sorbitol   0.33 M
MgCl$_2$ 0.005 M
Na$_4$P$_2$O$_7$.10H$_2$O   0.01 M

Adjust pH of above mixture to 6.5 with HCl and add sodium isoascorbate to a final concentration of 0.002 M just before use.

Re-suspending medium:

Sorbitol	0.33 M
$MgCl_2$	0.001 M
$MnCl_2$	0.001 M
EDTA (ethylenediamine tetra acetate)	0.002 M
HEPES (hydroxyethylpiperazine-ethanesulphonic acid – buffer)	0.05 M

Adjust pH to 7.6 with NaOH.

Homogenize 50 g of chilled spinach leaves with 200 cm^3 of freshly made grinding medium for 3 to 5 s in a domestic blender. Squeeze the macerate through two layers of cheesecloth and filter through eight layers of cheesecloth into 50-cm^3 plastic centrifuge tubes. Centrifuge rapidly at 0°C from rest to 4000 g to rest in approximately 90 s. Re-suspend the pellet gently using a glass rod and a small piece of absorbent cotton in 1 cm^3 of re-suspending medium. This procedure should produce a suspension of chloroplasts, 50–80% Type A (complete), which would be capable of high rates of CO_2 fixation.

Notes   The yield of chloroplasts is increased if the leaves are floated on cold running water or in an ice bath and brightly illuminated for about 30 min prior to grinding. To avoid too much starch in the leaves, spinach should be harvested early in the morning – the presence of white rings in the chloroplast pellet is indicative of excessive starch and calcium oxalate in the spinach and results in the disruption of the chloroplasts. A number of species other than *Spinacea oleracea* are often called spinach (e.g. beet and Swiss chard); however, these do not yield chloroplasts with good CO_2 fixation rates. Peas, lettuce and *Chenopodium album* are acceptable substitutes for spinach.

3.2   Chloroplast pigments

All photosynthetic organisms contain one or more organic pigments capable of absorbing visible radiation which will initiate the photochemical reactions of photosynthesis. These pigments can be extracted from most leaves into alcohol or into other organic solvents. From the alcoholic extract, individual pigments can be separated by chromatography on a column of powdered sugar, as was shown by the Russian botanist Tswett in

Table 3.1. *The photosynthetic pigments*

Type of pigment	Chararacteristic absorption maxima (nm) (in organic solvents)	Occurrence
Chlorophylls		
Chlorophyll *a*	420,660	All higher plants and algae
Chlorophyll *b*	435,643	All higher plants and green algae
Chlorophyll *c*	445,625	Diatoms and brown algae
Chlorophyll *d*	450,690	Red algae
Carotenoids		
β-carotene	425, 450, 480	Higher plants and most algae
α-carotene	420, 440, 470	Most plants and some algae
Luteol	425, 445, 475	Green algae, red algae and higher plants
Violaxanthol	425, 450, 475	Higher plants
Fucoxanthol	425, 450, 475	Diatoms and brown algae
Phycobilins		
Phycoerythrins	490, 546, 576	Red algae and in some cyanobacteria
Phycocyanins	618	Cyanobacteria and in some red algae
Allophycocyanins	650	Cyanobacteria and red algae

1906. The three major classes of pigments found in plants and algae are the chlorophylls, the carotenoids and the phycobilins. The chlorophylls and carotenoids are insoluble in water but the phycobilins are soluble in water. The carotenoids and phycobilins are called the accessory photosynthetic pigments since the quanta absorbed by these pigments can be transferred to chlorophyll. Table 3.1 gives the absorption characteristics of these pigments. The photosynthetic pigments of bacteria are discussed in Chapter 7.

Chlorophylls are the pigments that give plants their characteristic green colour. They constitute about 4% of chloroplast dry mass. They are insoluble in water but soluble in organic solvents. Chlorophyll *a* is bluish-green and chlorophyll *b* is yellowish-green. Chlorophyll *a* is present in all photosynthetic organisms which evolve O_2. Chlorophyll *b* is present (about one-third of the content of chlorophyll *a*) in leaves of higher plants and in green algae. They are absent in red algae and cyanobacteria. The absorption

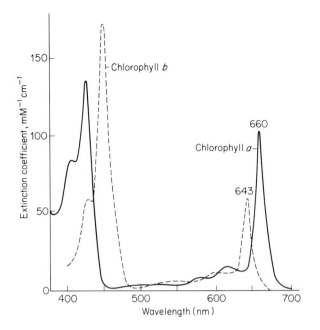

Fig. 3.6 Absorption spectra of chlorophylls extracted in ether.

maxima of chlorophyll *a* and chlorophyll *b* in ether are respectively at 660 and 643 nm, as shown in Fig. 3.6; in acetone the peaks are at 663 and 645 nm. However, careful spectroscopic investigations of the living cell indicate the presence of multiple forms of chlorophyll *a in vivo*. These forms of chlorophyll *a* may be associated in different ways with the lamellae and have different photochemical functions.

Systematic studies by Willstatter, and later by Fischer (both German Nobel Laureates), established the molecular structure of chlorophyll. The molecular formula for Chl *a* is $C_{55}H_{72}N_4O_5Mg$ ($M_r = 892$) and for Chl *b* is $C_{55}H_{70}N_4O_6Mg$ ($M_r = 906$). The structural formula for chlorophyll determined by Fischer in 1940 from degradative studies was confirmed by the complete synthesis of the molecule by Woodward at Harvard in 1960 (Nobel Prize 1965). The chlorophyll molecule (Fig. 3.7) contains a porphyrin 'head' and a phytol 'tail'. The polar (water soluble) porphyrin nucleus is made up of a tetrapyrrole ring and a magnesium atom. In the cell, electron microscopists think that the chlorophyll is sandwiched between protein and lipid layers of the chloroplast lamellae. The porphyrin part of the molecule is bound to the protein while the phytol chain extends into the lipid layer since it is soluble in lipids. Pheophytins, which are chlorophylls

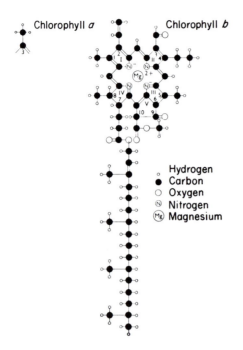

Fig. 3.7   Chlorophyll *a* and *b* structures.

without the central Mg atom, have been identified as constituents of the photosynthetic electron transport chain.

Estimation of chlorophyll   The optical absorption curves of chlorophyll *a* and chlorophyll *b*, in acetone solution, intersect at 652 nm. A solution of chlorophyll in acetone at a concentration of 1 mg cm^{-3} has an absorbance of 34.5 at 652 nm. The following method can be used for the determination of the total chlorophyll of a chloroplast preparation. Dilute 0.1 cm^3 of the chloroplast suspension to 20 cm^3 (200 fold) with 80% acetone, mix well, and filter. Measure the absorbance (A) of the filtrate in a 1-cm light path cell against 80% acetone as reference.

$$\text{Chl (mg cm}^{-3}) \text{ in the original suspension} = \text{A652} \times \frac{200}{34.5} = \text{A652} \times 5.8$$

For determination of the chlorophyll content of leaves the simplest method is to freeze the leaf in liquid N$_2$, grind it to a powder, mix with 80% acetone in the dark and then centrifuge or filter. Measure the absorbance

(A) of the acetone extract at 664 and 647 nm in an 1-cm path length cuvette. The following equations are used to calculate Chl a, Chl b and the total chlorophyll.

Chl a (μM) = 13.19 A664 − 2.57 A647.
Chl b (μM) = 22.10 A647 − 5.26 A664.
Total chlorophyll (μM) = 7.93 A664 + 19.53 A647

These equations are derived from the known extinction coefficients of Chl a and Chl b at 664 and 647 nm. The molecular weights of Chl a and Chl b are 892 and 906 kDa, respectively.

Cyanobacteria, which contain only Chl a, do not release their chlorophyll readily into solution. For chlorophyll determination, the cells are incubated in 90% (v/v) methanol for 1–6 hours at 4°C in the dark and then centrifuged. The absorbance of the methanol extract is measured at 665 nm. The Chl a content of the cells is calculated using the equation: Chl a (μg cm^{-3}) = A665 × 13.9. (The extinction coefficient ϵ of Chl a in methanol, ϵ 665 = 72 mg^{-1} l^{-2}).

Chl a concentration of an 80% acetone extract can be calculated from A663 using the formula:

Chl a (μg cm^{-3}) = A663 × 12.2 (ϵ663 = 82 mg^{-1} Chl a l^{-1})

The carotenoids are yellow or orange pigments found in all photosynthesizing cells. Their colour in the leaves is normally masked by chlorophyll, but in the autumn season when chlorophyll disintegrates the yellow pigments become visible. Carotenoids contain a conjugated double bond system of the polyene type. They are usually either hydrocarbons (carotenes) or oxygenated hydrocarbons (carotenols or xanthophylls) of 40 carbon chains built up from isoprene subunits (Fig. 3.8). They have triple-banded absorption spectra in the region from about 400 to 550 nm. The carotenoids are situated in the chloroplast lamellae in close proximity to the chlorophyll. The energy absorbed by the carotenoids may be transferred to chlorophyll a for photosynthesis. In addition the carotenoids may protect the chlorophyll molecules from too much photo-oxidation in excessive light.

Blue-green algae and red marine algae contain a group of pigments known as *phycobilins* (Fig. 3.9). Phycobilins are linear tetrapyrroles structurally related to chlorophyll a but they do not have the phytyl side chain, nor do they contain magnesium. The chromophores of phycobilins are covalently linked to polypeptides to form water-soluble phycobiliproteins.

In the intact cyanobacterial cells (and in red algal chloroplasts) the

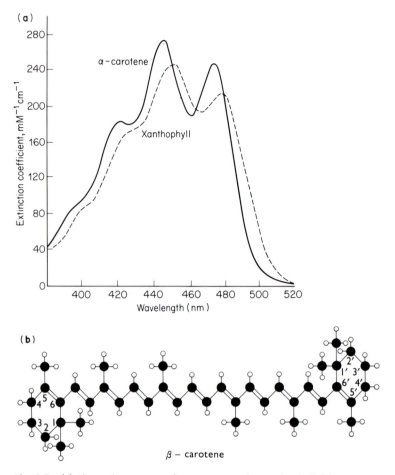

Fig. 3.8   (**a**) Absorption spectra of α-carotene and a xanthophyll. (**b**) Structure of β-carotene.

phycobiliproteins are assembled into multimeric particles named *phycobilisomes* which are attached in regularly spaced rows to the protoplasmic surface of the thylakoid membrane. They are the main constituents of the LHCII antenna of these organisms. Phycobilisome morphology and composition vary with the organism of its origin. The ultrastructure of the phycobilisomes has been studied by freeze-fracture and transmission electron micrography and recently by X-ray crystallographic analysis. In most cyanobacteria they consist of a bicylindrical or tricylindrical core with six peripheral rods radiating from it in a hemidiscoidal arrangement (Fig. 3.10). Allophycocyanin is the major component of the core complex. The

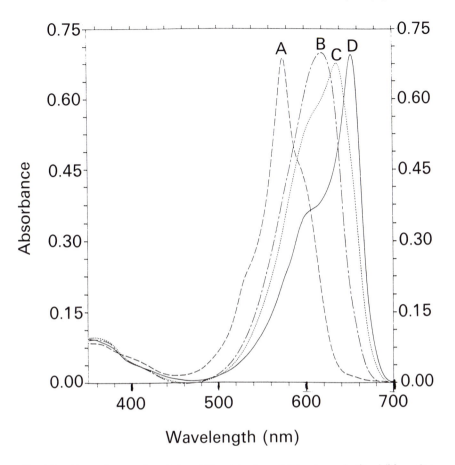

Fig. 3.9 Absorption spectra of phycobiliproteins from *M. laminosus* in the visible region. **A.** Phycoerythrocyanin 575 nm; **B.** phycocyanin trimer $(\alpha\beta)_3$ 620 nm; **C.** phycocyanin hexamer $(\alpha\beta)_6$ 635 nm; **D.** allophycocyanin 652 nm. (Courtesy: W. Sidler, ETH, Zurich.)

portion of the rods adjacent to the core is made up of phycocyanin: phycoerythrin or phycoerythrocyanin, when present, is found at the periphery of the rods. In addition to the phycobiliproteins which absorb visible light of various wavelengths, the phycobilisomes also contain colourless *linker polypeptides* which function in the assembly of the phycobilisomes.

Each phycobilisome is in contact with two PSII reaction centres. The energy absorbed by the particles is funneled into the PSII reaction centre by radiationless pathways.

The red phycoerythrins found in all red algae (see Table 3.1) absorb light

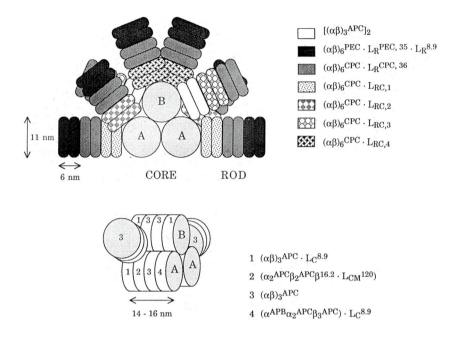

Legend:

$[(\alpha\beta)_3^{APC}]_2$

$(\alpha\beta)_6^{PEC} \cdot L_R^{PEC, 35} \cdot L_R^{8.9}$

$(\alpha\beta)_6^{CPC} \cdot L_R^{CPC, 36}$

$(\alpha\beta)_6^{CPC} \cdot L_{RC,1}$

$(\alpha\beta)_6^{CPC} \cdot L_{RC,2}$

$(\alpha\beta)_6^{CPC} \cdot L_{RC,3}$

$(\alpha\beta)_6^{CPC} \cdot L_{RC,4}$

11 nm

6 nm CORE ROD

14 - 16 nm

1 $(\alpha\beta)_3^{APC} \cdot L_C^{8.9}$

2 $(\alpha_2^{APC}\beta_2^{APC}\beta^{16.2} \cdot L_{CM}^{120})$

3 $(\alpha\beta)_3^{APC}$

4 $(\alpha^{APB}\alpha_2^{APC}\beta_3^{APC}) \cdot L_C^{8.9}$

Fig. 3.10 Proposed schematic model of the hemidiscoidal phycobilisome of *Mastigocladus laminosus*. The core consists of three cylinders and two hexamer equivalents of APC forming the base of two rods. The abbreviations APC, APB, CPC and PEC are used for the phycobiliproteins allophycocyanin, allophycocyanin B, phycocyanin, and phycoerythrocyanin respectively, and α^{APC}, β^{APC} etc. for the α and β subunits of these proteins. Linker polypeptides are abbreviated with L, with the superscript denoting the apparent size in kDa and the subscript specifying the location of the polypeptide. R, peripheral rod substructure; RC (1–4) one of four rod–core junctions; CM, core–membrane junction. (Courtesy: W. Sidler.)

in the middle of the visible spectrum. This enables the red algae living under the sea to perform photosynthesis in the dim bluish-green light reaching the lower surfaces of the ocean – the deeper under the sea a red alga lives the more phycoerythrin it contains in relation to chlorophyll. The blue phycocyanins and allophycocyanins occur in the blue-green algae which live on the surface layers of lakes and on land.

The energy transfer in the red alga, *Porphyridium cruentum*, occurs in the sequence:

phycoerythrin→phycocyanin→allophycocyanin→chlorophyll *a*.

Thus higher plants and algae, during the course of evolution, have developed various pigments to capture the available solar radiation most

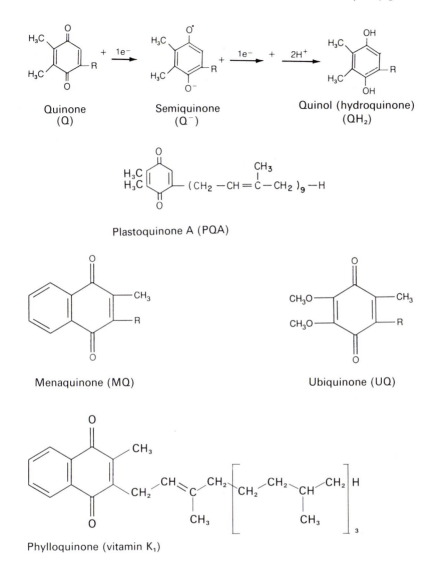

Quinone
(Q)

Semiquinone
(Q⁻)

Quinol (hydroquinone)
(QH₂)

Plastoquinone A (PQA)

Menaquinone (MQ)

Ubiquinone (UQ)

Phylloquinone (vitamin K₁)

Fig. 3.11   The chemical structure of some of the quinone derivatives found in photosynthetic complexes of chloroplasts and bacteria.

efficiently and to carry out photosynthesis. The relative abundance of these pigments depends upon the species, the location of the plant, the seasons, etc.

In addition to the conjugated pigment–protein complexes which are involved in light harvesting and energy transduction, the chloroplast

Table 3.2. *Properties of chloroplast electron transfer chain components* (*see also* Fig. 5.2, §5.2 and 5.3)

Component and symbol	Molecular mass (daltons)	E_m; mid-point redox potential (volts) (see §5.1)	Detection, probable function, etc.
'Water-oxidizing' complex 'M' (hypothetical)		+0.82?	EPR, EXAFS, reconstitution studies, cyanobacterial mutants. Probably contains 4 Mn ions and Ca^{2+} and Cl^-. Oxidizes H_2O. Probable ligands are EP33 and D_1–D_2 proteins donating electrons to Y_Z and releasing protons into the lumen of thylakoid.
Primary e^- donor to PSII reaction centre Y_Z		+1.0	EPR, site-directed mutagenesis; a tyrosine bound to D_1 protein of the PSII reaction centre (RC); $1e^-$ mediator between 'M' and P680.
P680	1 or 2 × 892	+1.0 −1.0 (excited state)	Flash absorption spectroscopy and EPR; special Chl a monomer or dimer bound to the D_1–D_2 protein or to 47 kDa RC protein; energy trap of PSII.
Pheophytin (Pheo)	868	−0.614	EPR; metastable intermediate e^- carrier from P*680 to Q_A.
Plastoquinones Q_A	740 (PQ A)	−0.15	UV absorption or fluorescence; first PSII stable e^- acceptor; $1e^-$ mediator between pheo and Q_B; may be bound to the D_2 protein.
Q_B			Flash absorption; two e^- gate between Q_A and PQ pool; reduced to semiquinone and then to quinol; semiquinone form binds to D_1.

PSII complex; predominantly in the grana (oppressed membranes)

Component	Molecular weight	Redox potential (V)	Method / function
Cyt b–f complex in grana and stroma lamellae			
Cytochrome b_{559}	9000 & 4000 Two subunits	+0.08 low potential / +0.38 high potential	Absorption spectroscopy; may be involved in O_2 evolution, cyclic e^- transport around PSII and photoprotection.
Cytochrome b_{563} (or b_6)	23 440	−0.05 high potential / −0.17 low potential	Absorption spectroscopy; $PQ \rightarrow PQH_2$ redox (Q) cycle ie. energy transduction; cyclic e^- transport; may be in protein phosphorylation.
Cytochrome f	34 000	+ ~0.35	Absorption spectoscopy; c-type cytochrome; $1e^-$ donor to PC.
Rieske iron–sulphur protein $[Fe–S]_R$	20 000	+0.29	EPR; $1e^-$ acceptor from PQ pool; Q cycle. e^- donor to cyt f.
PSI complex; predominantly in the stroma lamellae (non-appressed membranes)			
P700	2 × 892	+0.48	Optical absorption, EPR; Chl a dimer bound to PSI A–B proteins; energy trap for PSI.
A_0	not known	−1.0	Flash EPR; metastable primary e^- acceptor from P700*; may be a Chl a monomer.
A_1	not known	not determined	Flash EPR, reconstitution studies; transient $1e^-$ mediator between A_0 and, most likely a phylloquinone (vitamin k_1).
F_X	not known	~ −0.73	EPR, redox titrations; $1e^-$ mediator to F_A and F_B; $[Fe–S]_4$ centre bound to PSI A–B proteins.
F_A / F_B	not known / not known	−0.55 / −0.59	EPR, redox titrations; $1e^-$ mediators between F_X and Fd; Two $[4Fe–4S]$ centres bound to PSI–C protein.
Ferredoxin – NADP reductase (FNR)	40 000	−0.38	Isolation from outer lamellae; 1 FAD/mole; e^- mediator between Fd and NADP; can act as diaphorase and transhydrogenase.

Table 3.2. (*cont.*)

Component and symbol	Molecular mass (daltons)	E_m; mid-point redox potential (volts) (see §5.1)	Detection, probable function, etc.
Plastoquinones (PQ)	Variable	~ -0.100	Optical absorption; abundant in chloroplasts; shuttles e^- and H^+ between PSII and cyt b–f complex; energy transduction and cyclic e^- transport from reduced Fd.
Plastocyanin (PC)	10 500 per Cu	$+0.38$	Isolation from leaves; 4 or 2 Cu per mole; lumen side of thylakoids; $1e^-$ mediator between cyt f and P700.
Ferredoxin (Fd)	11 000	-0.42	EPR, optical spectrum; easily isolated from leaves and algae; soluble non-haem iron protein with [2Fe–2S] centre; cyclic and non-cyclic e^- transport; $1e^-$ donor to various chloroplast constituents.

Mobile e^- carriers

lamellae also contain many other proteins, lipids, quinones, and ions. The properties of these and other chloroplast electron transport chain components are summarized in Table 3.2 and will be discussed in more detail in later chapters.

Plastoquinones (quinones in plastids or chloroplasts) are the most abundant electron mediators found in chloroplasts. Plastoquinones possess a dimethyl benzoquinone ring to which long-chain aliphatic hydrocarbons (phytyl) are attached. The structures of plastoquinone A and some other quinones occurring in photosynthetic reaction centres are given in Fig. 3.11. The unsubstituted quinone, benzoquinone, and dimethyl benzoquinone are used as artificial electron acceptors (Hill reagents) of photosystem II. The quinone ring can be reduced to a semiquinone free radical (addition of an electron) or to a quinol (addition of two protons and electrons).

The *lipids* play a dynamic role in the function of the thylakoid membrane components. The more abundant lipids in the bilayer matrix of the membrane are the glycolipids, monogalactosyl diacylglycerol and digalactosyl diacylglycerol, which constitute about 75% of the membrane lipids. Recent studies using tritium labelling (exchange of H atoms with ^3H) of right-side-out and inside-out membrane vesicles (see Chapter 4) indicate an asymmetric distribution of these lipids – about 60% of both galactolipids located in the outer half and the rest in the inner half of the bilayer. The minor lipid components – the sulpholipids and phospholipids – are mainly distributed in the inner half of the bilayer.

3.3   The photosynthetic unit

Functionally, chlorophyll molecules act in groups. A photosynthetic unit was conceived as a group of pigments and other molecules utilizing the transfer of excitation energy as a mechanism by which the reaction centre communicated with an antenna of light-harvesting pigments, as shown in Fig. 3.12. According to this concept, a single quantum of energy absorbed anywhere in a set of about 250 to 300 chlorophyll molecules migrates to a reaction centre containing a special pair of chlorophyll *a* molecules and promotes an electron transfer event. The following observations led to the idea of the existence of a photosynthetic unit.

1. Approximately 8 quanta of light absorbed by chlorophyll are required for the photosynthetic reduction of 1 CO_2 molecule and the evolution of 1 O_2 molecule. If each chlorophyll molecule in an algal cell can react

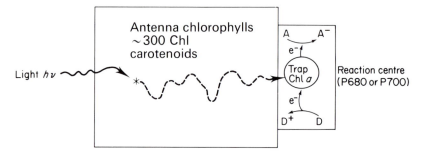

Fig. 3.12   Diagrammatic representation of a photosynthetic unit in chloroplasts. A photon captured by an antenna chlorophyll is transferred to a trap chlorophyll *a*, where photochemical charge separation takes place.

photochemically, a sufficiently intense flash of light should bring about the evolution of 1 O_2 for every 8 chlorophyll molecules present. However, the flashing light experiments of Emerson and Arnold (Chapter 2) on *Chlorella* suspensions showed that the maximum yield per flash was 1 O_2 molecule for about 2500 chlorophyll molecules, that is, a quantum of light is absorbed by one in a cluster of about 300 chlorophyll molecules.

2. Gaffron and Wohl, in Chicago, calculated that a single chlorophyll molecule in a dimly illuminated plant will absorb a light quantum only once in several minutes. At this rate a single molecule of chlorophyll will require nearly an hour to capture the light quanta needed for the evolution of one molecule of O_2. But when a plant is illuminated, the maximum rates of CO_2 uptake and O_2 evolution are quickly established. So Gaffron and Wohl postulated that the energy harvested by a large set of chlorophyll molecules is conducted to a single reaction centre.

3. Gaffron and co-workers also observed that certain golden-yellow leaves of tobacco, with very little chlorophyll, may reach a nearly normal rate of photosynthesis at very high light intensities. These leaves would have contained a larger proportion of the special type of chlorophyll molecule which is in direct contact with the components of the electron transfer chain.

4. Difference spectroscopy has revealed the role of certain special components, like P700 (by Kok in 1956) and cytochrome (by Duysens in 1961), in the photochemical electron transfer reactions. There is 1 molecule of light-reacting cytochrome and 1 P700 for every 250 chlorophyll molecules in higher plants and algae.

This concept of a photosynthetic unit is now of historical interest only. It has been revised a number of times in the last 30 years, in the light of

Table 3.3. *Morphological cascade of plant structures and function (from J. Goudriaan).*

Organ or organelle	Average dimension	Process	Relaxation time
Chlorophyll	1 nm	Excitation	10^{-15} s
Photosynthetic units	20 nm	Fluorescence	10^{-9} s
Thylakoid membrane	25 nm thick, 259 nm long	Ion leakage	0.1 s
Chloroplasts	1 μm thick, 10 μm long	Calvin cycle	100 s
Mesophyll cells	20 μm	Glucose and starch synthesis	1 day
Leaf	0.3 mm thick, 10 cm wide	Lifetime	10 days
Whole plant	0.2–10 m height	Lifetime	100 days

important discoveries and postulates, and is sure to be modified again to accommodate data emerging from new experimentation. Some of these developments include (*a*) the observation that two photosystems with different reaction centre pigments operate in synchrony in the chloroplasts, as proposed in the Z scheme by Hill and Bendall; (*b*) the discovery of the association of O_2 evolution with photosystem II and of ferredoxin reduction with photosystem I; (*c*) the isolation of reaction centres and different membrane complexes from the photosynthetic membranes and the localization of photosystems and electron transfer chain components in the membranes as a result of electron microscopic and immunological cross-reaction studies.

A photosynthetic unit is presently conceived as an integrated assembly of about 500 chlorophyll molecules (250 to 300 per reaction centre) and an electron transport chain that could independently harvest light, resulting in O_2 evolution and NADP reduction.

The dimensions and properties of photoactive components from an ordinary plant to a chlorophyll molecule are given in Table 3.3.

3.4   Photosynthetic apparatus of C₄ plants

Leaves of plants like sugar cane, maize (corn), *Sorghum*, *Amaranthus* and many tropical grasses contain two distinct types of chloroplasts. The leaves of these plants possess 'Kranz-type' anatomy (kranz in German means

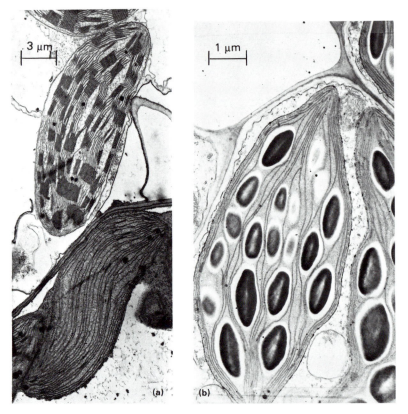

Fig. 3.13 (a) Two different types of chloroplasts in the maize leaf (a C_4 plant). **Above,** mesophyll or granal type. **Below,** bundle-sheath or agranal type. There are no starch grains in the bundle-sheath chloroplast as the leaf was kept in darkness for 24 hours before fixation. (Courtesy: G. Montes.) **(b)** Bundle-sheath (agranal) chloroplast from maize with deposits of starch granules. (Courtesy: J. M. Whatley and F. R. Whatley, Oxford.)

wreath). The chloroplasts are located in the leaf cells in two concentric layers surrounding the vascular bundle; the inner layer is called the *bundle-sheath* and the outer the *mesophyll*. Chloroplasts from these plants possess a membrane system in the peripheral part of the stroma, called the *peripheral reticulum*, which connects the thylakoid membranes to the chloroplast envelope. The two types of chloroplasts can be separated by careful grinding and centrifugation using density gradients. As will be discussed in Chapter 6, these plants are able to fix CO_2 by two different pathways: (*a*) the normal Calvin cycle in the bundle-sheath chloroplasts where the initial product of CO_2 fixation is the three-carbon compound phosphoglyceric acid (C_3 pathway); and (*b*) in the mesophyll chloroplasts by combining CO_2 with phosphoenol pyruvate to produce the four-carbon acids oxaloacetate

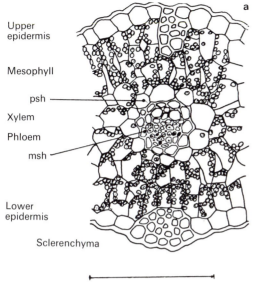

Upper
epidermis

Mesophyll

psh

Xylem

Phloem

msh

Lower
epidermis

Sclerenchyma

a

0.1 mm

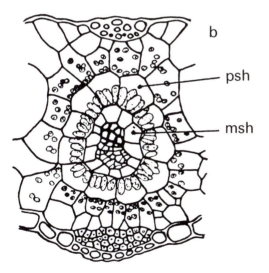

b

— psh

— msh

Fig. 3.14 Cross-sections of leaves from different types of plants: (**a**) C₃ species; (**b**) and (**c**) C₄ species (psh = parenchyma bundle-sheath, msh = mesotome bundle-sheath); (**d**) cross-section of a leaf from a 'sunplant', *Fagus silvatica*. (Courtesy: H. R. Bolhar-Nordenkampf, University of Vienna.)

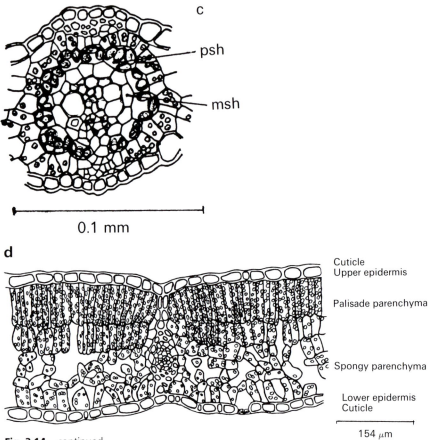

c

psh

msh

0.1 mm

d

Cuticle
Upper epidermis

Palisade parenchyma

Spongy parenchyma

Lower epidermis
Cuticle

154 μm

Fig. 3.14   continued.

and malate (C_4 pathway). Plants which fix CO_2 by the Calvin cycle only are termed C_3 species and are generally grown in the temperate zones, e.g. wheat, spinach, oak tree. Those plants which fix CO_2 in the light both by the Calvin cycle and by the C_4 acid pathway are termed C_4 species and are generally adapted to growth in warmer and/or drier climates.

The mesophyll chloroplasts of C_4 plants (Fig. 3.13a) are randomly distributed in the cell, have stacks of grana and few starch granules. The bundle-sheath chloroplasts are relatively larger in size, generally lack grana and possess a number of starch granules (Fig. 3.13b). The mesophyll and bundle-sheath chloroplasts lie adjacent in the leaf and photosynthetic products can easily flow from one type to the other (Fig. 3.14). Generally, plants possessing these dimorphic-type chloroplasts have a very low CO_2 compensation point, very low photorespiration and glycollate metabolism, and grow more rapidly with higher crop yields (see Chapter 6).

4

Light absorption and the two photosystems

The most stable states of atoms are those in which the valence electrons are distributed, in accordance with the Pauli principle, into the quantum states of least energy, i.e. the electrons are in their ground states of energy level. When light is absorbed by an atom in the ground state, the whole energy of the quantum (hv) is added to it, and the electrons are lifted to an energy-rich excited state. This is illustrated in Fig. 4.1, taking the helium atom as an example. The time needed for the entire process is of the order of 10^{-15} seconds. If an atom has an even number of electrons, the spins of these electrons are usually arranged in opposite directions and cancel each other out so that the total electronic spin of the atom $s = 0$ (singlet state). For an atom with an odd number of electrons the net spin is $s = \frac{1}{2}$ (doublet state). If an atom has an even number of electrons and if the electronic spins are in the parallel direction, the net spin is $s = 1$ and the atom is said to be in the triplet state. These states are illustrated in Fig. 4.1. The transition from the ground to the excited state of an atom by the absorption of a single quantum of

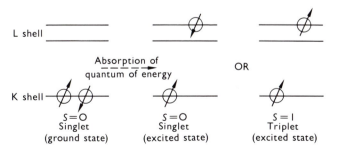

Fig. 4.1 Energy levels of electrons in the helium atom.

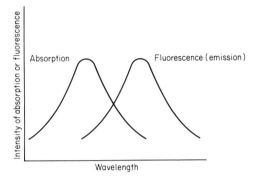

Fig. 4.2 Spectra of light absorption and emission.

energy can be followed by a sharp line in its absorption spectrum at the wavelength λ given by $\Delta E = hc/\lambda$ (see Chapter 1). In a molecule consisting of various atoms, the transition from the ground to the excited state can take place by the absorption of light of varying amounts of energy quanta; the sharp line of the atomic absorption spectrum then is replaced by a broad absorption band. In individual atoms absorption and emission take place at the same wavelength, while in a whole molecule the absorption and emission spectra do not coincide; the peak of the emission spectrum is at a longer wavelength than the peak of the corresponding absorption spectrum (Fig. 4.2).

4.1 Dissipation of absorbed light energy: photochemistry, fluorescence, phosphorescence and thermoluminescence

A molecule of chlorophyll on the absorption of light becomes excited to a higher energy state, absorption of a blue photon (2.5 eV) raising the energy level to state 2 and that of a red photon with less energy (1.8 eV) to the excited state 1. An excited molecule is not stable and the electrons return rapidly to their ground level, releasing the absorbed photon energy in a number of ways (Figs. 4.3 and 4.4): (1) the electronic excitation energy can be transferred to another acceptor molecule; this results in photosynthetic electron transport; (2) it can release part of the excitation as heat (thermal dissipation); and (3) it can emit the rest of the energy as a photon of lower energy content (i.e. higher wavelength); this phenomenon is called *fluorescence*. The emission maximum of the fluorescence spectrum of a molecule is

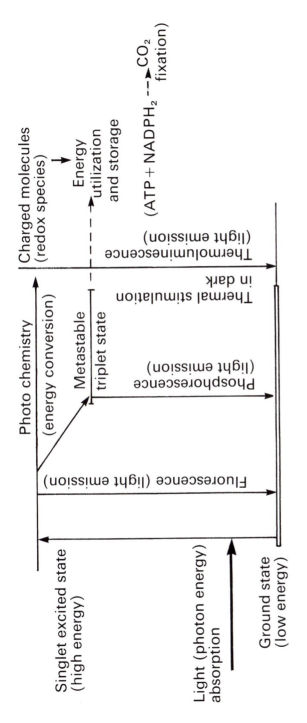

Fig. 4.3 Schematic illustration of light absorption and energy dissipation by photosynthetic materials.

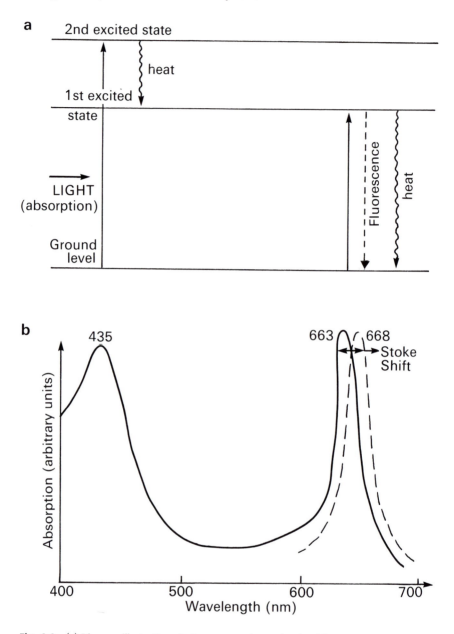

Fig. 4.4 (**a**) Diagram illustrating photon energy absorption by chlorophyll and the non-photochemical dissipation of a fraction of the absorbed energy as heat and as fluorescence. (**b**) Absorption (bold line) and fluorescence spectra of chlorophyll showing the Stoke shift.

always at longer wavelength than the wavelength of the corresponding absorption spectrum. Chlorophyll *a* extracts, for example, absorb in the blue and red regions of the spectrum but fluoresce only in the red. This is due to the fact that the decay of excitation from state 2 to state 1 occurs within 10^{-13} s of photon capture, whereas fluorescence discharge requires about 10^{-9} s. For this reason chlorophyll fluorescence always occurs via excited state 1 in the red, irrespective of the quality of exciting light (see Plate V). The average time a molecule spends in the metastable state between absorption and emission of radiation is termed the natural lifetime of the excited state; the average fluorescence lifetime is of the order of 5–7 nano (10^{-9}) seconds.

The maximum intensity of fluorescence for a Chl *a* molecule in solution occurs at 668 nm, compared to the maximum absorption at 663 nm; the red shift in the absorption maximum is called *Stoke's shift* (Fig. 4.4). In isolated chloroplasts and in intact leaves and algae, the wavelength of maximum fluorescence measured is around 685 nm and is mainly contributed by the light-harvesting complexes of photosystem II. At low temperature (liquid N_2; 77 K, i.e. $-196°C$) the fluorescence yield increases due to contributions from the photosystem II reaction centre core and photosystem I antenna chlorophylls (Fig. 4.5).

Fluorescence emission characteristics can be conveniently monitored without damage to the photosynthetic material and thus the technique is being extensively used in photosynthesis research (§8.12), both in the laboratory and in the field.

A fourth route by which an excited molecule can lose its energy is by transfer from its original excited singlet state into a metastable triplet state with a much longer lifetime (of the order of milliseconds) by a mechanism called intersystem crossing. From the metastable triplet state the molecule can revert to the natural ground state by emitting a photon at a longer wavelength. This weak emission is known as *phosphorescence*. Phosphorescence is slow enough to be observed by the eye even when the exciting light is turned off. Due to their longer lifetime, lower energy, and magnetic moment (since the excited electron and its partner have parallel spins), the *triplet excited states* can be of importance in photochemistry. As yet there is no conclusive evidence for the participation of chlorophyll triplet states in photosynthesis, although their formation has been detected in PSII under high light.

The primary event in all photosynthetic processes is the light-induced transfer of an excited electron from a donor species D to a closely bound

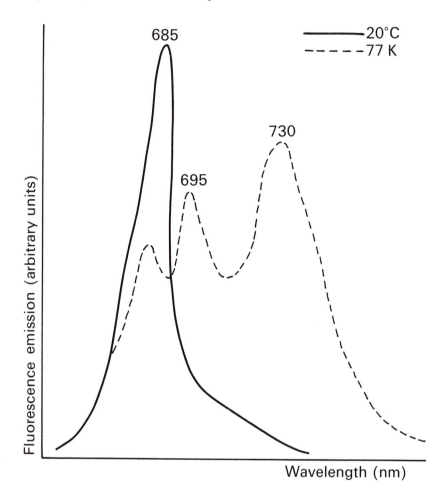

Fig. 4.5 Fluorescence spectra of chloroplasts at 20°C and at 77 K (-196°C, liquid N_2). The chlorophylls contributing to the fluorescence are PSII antennae (685 nm), PSII RC core (695 nm), and PSI (730 nm). (See also Fig. 8.5.)

acceptor A. In chloroplasts, D is a special type of chlorophyll conjugated to a protein and located in the *reaction centre* (§4.5). Thus:

$$DA \xrightarrow{h\nu} D^*A \longrightarrow D^+A^-$$

where D* is a chlorophyll molecule which has acquired an exciton (quantum of excited energy). The energy trapped in the charge separation is subsequently utilized for photosynthetic electron transport – thus light energy is converted to chemical energy. Antenna pigments transfer their absorbed electronic excitations to the reaction centre as singlet excitons, and the reaction centre photochemistry is also initiated from an excited singlet state. The initial charge separation event happens within a time span of a few picoseconds after the capture of a photon.

When chloroplasts are illuminated at room temperature and immediately kept in the dark they are able to emit light, the emission resembling that of chlorophyll fluorescence. This light emission, observable from microseconds to minutes after terminating illumination, is called *delayed fluorescence* or *delayed light emission*. Delayed fluorescence is caused by the spontaneous recombination of some of the charged species formed in the reaction centre $(D^+ A^-)$ to regenerate D^*, and release of chemical energy as light. The intensity of delayed fluorescence is a reflection of the energized state (proton gradient) of the thylakoid membranes and of the rate of electron transfer from A^- to subsequent components of the electron transport chain. Delayed light emission can also occur by a reversal of the energy transition from the excited singlet state to the excited triplet state. The exact mechanism of the process is still not fully understood.

Thermoluminescence   If leaves, algal cells or chloroplasts are frozen rapidly after illumination (or illuminated at low temperatures, e.g. at 77 K in liquid N_2) and subsequently warmed in the dark, sudden outbursts of light are observed at certain characteristic temperatures. This phenomenon is referred to as *thermoluminescence*, which is a special type of thermally stimulated delayed light emission. In the frozen plant material, the metastable charged species produced by the light-induced primary charge separation at photosystem II are stabilized. On thermal activation, these charged species recombine and release the energy of recombination as light.

Thermoluminescence originates mainly from PSII, and the light-emission bands occurring at various defined temperatures have been correlated with the functioning of the water-oxidizing complex, decay halftimes of charged species in PSII, herbicide binding to PSII, etc. Minor contributions to thermoluminescence are made by the light-harvesting complexes. The phenomenon is being extensively applied nowadays in the study of S-state transitions (§4.8) during O_2 evolution, and to a lesser extent in defining the role of Mn and Ca ions in O_2 evolution and in photoinhibition studies.

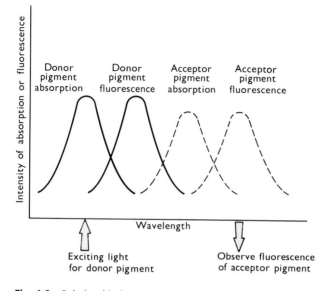

Fig. 4.6 Relationship between absorption and fluorescence spectra of donor and acceptor pigments illustrating sensitized fluorescence.

4.2 Energy transfer or sensitized fluorescence

The phenomenon of sensitized fluorescence involves the interaction of two molecules that may be separated in solution by many molecules of the solvent. In this type of energy transfer, two kinds of pigment are dissolved in the same solvent and the solution is illuminated with light of such wavelength that can be absorbed by only one of the pigments, called the *donor*. The wavelength of light emitted from the solution, however, corresponds to the fluorescence spectrum of the second pigment molecule, the *acceptor*. The energy of excitation of the donor molecule is transferred by resonance to the acceptor molecule. One of the requisites for this type of energy transfer is that the fluorescent state of the donor molecule must have an energy greater than or equal to the fluorescent state of the acceptor molecule, that is, the fluorescence band of the donor molecule should overlap with the absorption band of the acceptor (Fig. 4.6). The quanta taken up by accessory pigments in many blue-green algae are transferred either wholly or partly to chlorophyll *a* by sensitized fluorescence. In the green algae, chlorophyll *a* fluorescence is observed when light is absorbed by chlorophylls *a* and *b* and also by the carotenoids. Recent fluorescence and flash-induced absorption change data indicate that the transfer of excitation energy from carotenoid singlets to Chl *a* occurs within 250 to 500 femto (10^{-15}) seconds. Chlorophyll *a* has its fluorescence at the longest wavelength so that the migration of energy is always from the other excited chloroplast pigments to chlorophyll *a*.

From studies of chlorophyll *a* fluorescence and quantum yields of photosynthesis in *Chlorella*, it was shown that in this alga the transfer of excitation energy from chlorophyll *b* to *a* is 100% efficient, while the energy transfer from the carotenoids to chlorophyll *a* is only 40% efficient. The close assembly of various pigment molecules in the lamellae is necessary for efficient energy transfer by this type of inductive resonance.

4.3   Action spectra, quantum yield

A graph showing the rate of photosynthesis (measured as O_2 evolution or CO_2 fixation) by monochromatic light as a function of the wavelength of light is known as the *action spectrum* of photosynthesis. For photochemical reactions involving a single pigment, the action spectrum has the same general shape as the absorption spectrum of the pigment.

In a photosynthetic reaction, the ratio of the absorbed quanta utilized in photochemical conversions to the total quanta absorbed is known as the quantum yield, or quantum efficiency of that reaction. For example, if P molecules of O_2 per second are evolved from a system which absorbs I quanta of monochromatic radiation per second, then the ratio P/I, (Φ), is called the *quantum yield* or *quantum efficiency* of photosynthesis at that wavelength. The reciprocal of the quantum yield $[1/\Phi]$, which gives the number of quanta required to liberate 1 molecule of O_2, is usually called the *quantum requirement* of photosynthesis. Although values ranging from 4 to 12 have been reported for the quantum requirement by various workers in the past, the widely accepted value nowadays is 8 or 9. The maximum quantum yield can only be determined under low irradiance when photosynthesis is light limited and is strictly proportional to the incident PPFD (Fig. 4.7). The quantum yields usually measured are 'apparent' since they are calculated on incident and not absorbed light. Emerson found an average Φ for O_2 evolution of 0.104 from 660 to 680 nm for *Chlorella*.

The main decay processes for exciton energy in PSII are photochemistry (P), heat (H) and fluorescence (F) (§4.1). In isolated chloroplasts or intact leaves, some of the excitation energy may also be dissipated to PSI by energy transfer (T). The contributions by the four processes can be

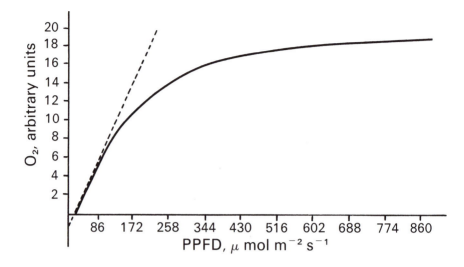

Fig. 4.7   Rate of photosynthesis (measured as O_2 evolution in an electrode) as a function of light intensity (PPFD). At first the rate increases linearly with increasing light. At relatively low PPFDs, the relationship departs from linearity due to metabolic constraints. The maximum quantum efficiency is calculated from the initial slope. Theoretically it should be 0.125 molecules of O_2 evolved per photon (a quantum requirement of 8). At a PPFD of 400 μmol m^{-2} s^{-1}, equal to one-fifth full sunlight, the rate of photosynthesis is already close to its maximum. (After Walker, 1992.)

expressed mathematically. Let us assume that there are n chlorophyll molecules and each molecule has a decay rate constant ks^{-1}. Then, the total de-excitations per second will be $n(k_p+k_H+k_F+k_T)$. If all the PSII reaction centres are in the 'open' state (i.e. ready to absorb photons), and if 'e' photons are absorbed per second by n chlorophylls, then the quantum yield of photochemistry,

$$\Phi_P = \frac{P}{en} = \frac{k_P}{(k_p + k_H + k_F + k_T)}$$

This value varies between 0.105 and 0.12 in leaves. Similarly, the quantum yield of fluorescence.

$$\Phi_F = \frac{F}{en} = \frac{k_F}{(k_p + k_H + k_F + k_T)}$$

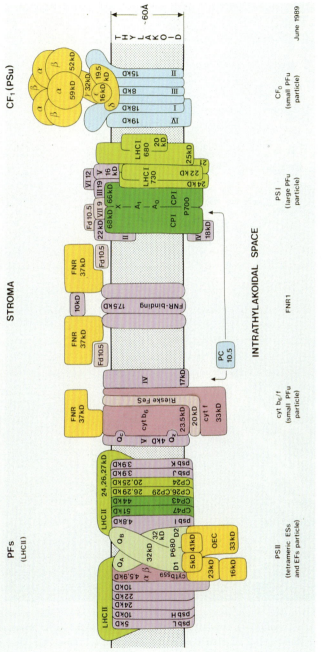

I Schematic arrangement of polypeptides which have been identified as components of barley thylakoid membranes from freeze-fracture electron microscopy, gel electrophoresis, and biochemical studies. CF$_0$ and CF$_1$, coupling factors; EF$_s$, endoplasmic fracture face of stacked thylakoids; ES$_s$, endoplasmic surface of stacked thylakoids; OEC, oxygen-evolving complex; PF$_s$, protoplasmic fracture face of stacked thylakoids; PF$_u$, protoplasmic fracture face of unstacked thylakoids. PS$_u$, protoplasmic surface of unstacked thylakoid. The molecular weight of some peptides may vary from species to species. (Courtesy: D. Simpson and D. von Wettstein, Carlsberg Laboratory, Copenhagen.)

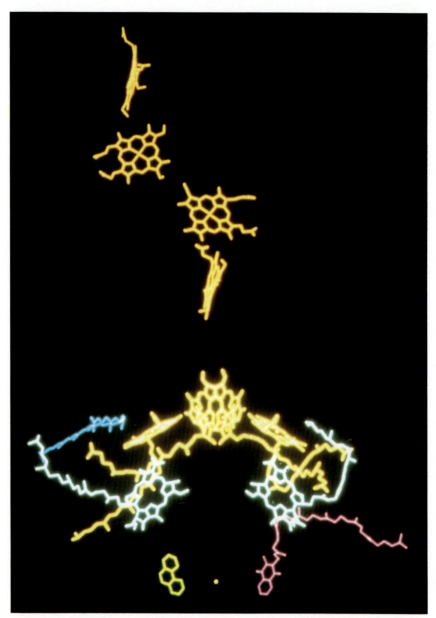

II Pigment arrangement in the photosynthetic reaction centre from *Rhodopseudomonas viridis* treated with orthophenanthroline (inhibitor of Q_B). 4 haem groups (light brown, top), 4 BChl molecules (yellow, centre), 2 BPheo molecules (light blue), 1 carotenoid (blue) and the primary acceptor Q_A can be seen. (Courtesy: H. Michel, Max-Planck Institute of Biochemistry, Frankfurt.)

III Chloroplasts in the guard cells of *Phyllitis scolopendrium*, a fern. (Courtesy: T. A. Mansfield, Lancaster University, Lancaster, UK.)

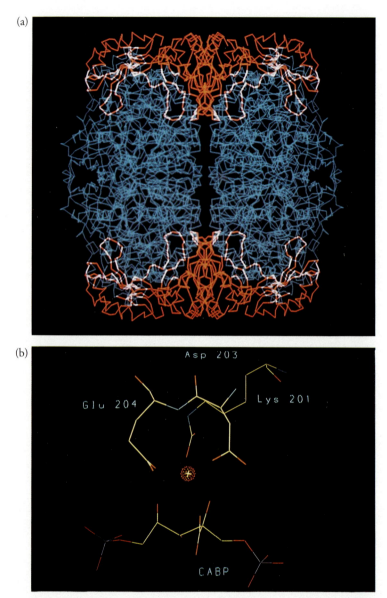

IV (a) The L_8S_8 arrangement of large (L, blue) and small (S, red) subunits in RuBisCO from spinach deduced from X-ray crystallography analysis. (b) The surroundings of the active site-Mg ion in the active site of spinach RuBisCO. CABP, 2-carboxy-D-arabinitol-1,-5-bisphosphate, is a transition state analogue in the carboxylation reaction, and the side-chain of Lys 201 is carboxylated by activator CO_2. (Courtesy: I. Andersson, C.-I. Bränden and S. Knight, Swedish Agricultural University, Uppsala.)

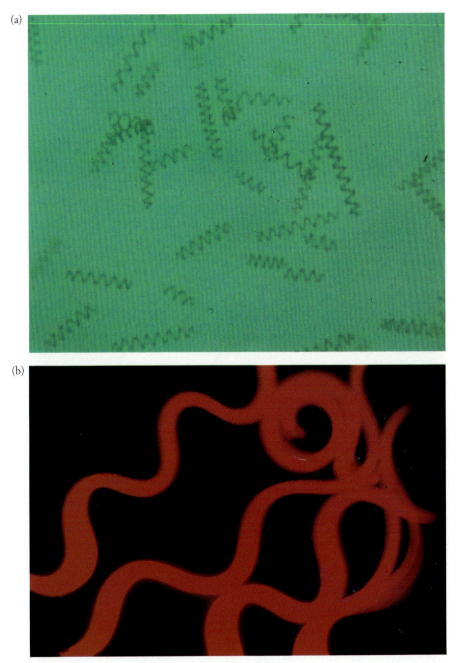

V (a) Cells of the cyanobacterium *Spirulina platensis*. (b) *S. platensis* fluorescing in light.
(Courtesy: A. Vonshak, Ben Gurion University, Israel.)

VI Portable LCA-4 IRGA for photosynthesis and transpiration measurements. (Courtesy: Analytical Development Co. Ltd (ADC), Hoddesdon, UK.)

VII The Hansatech Plant Efficiency Analyzer (PEA) for measuring photosynthesis by the fluorescence technique. (Courtesy: Hansatech Instruments Ltd, King's Lynn, Norfolk, UK.)

VIII Portable leaf area meter for measurements of areas of leaves in the field and laboratory.
(Courtesy: LI-COR, Inc., Lincoln, Nebraska, USA.)

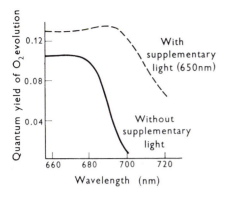

Fig. 4.8   Efficiency of photosynthesis (quantum yield) in the green alga *Chlorella* at different wavelengths of light. With supplementary light, the quantum yield is enhanced at wavelengths above 680 nm – the Emerson enhancement effect.

4.4   Emerson effect and the two light reactions

Emerson and associates, at the University of Illinois in the 1940s, studied the action spectra of photosynthesis for various algae by measuring the maximum quantum yield of photosynthesis as a function of the monochromatic light used to illuminate the algae. They found that the most effective light for photosynthesis, in *Chlorella*, was red (650 to 680 nm) and blue (400 to 460 nm), those colours that are most strongly absorbed by chlorophyll. The photosynthetic efficiency of a quantum absorbed at 680 nm was about 36% more than that of a quantum absorbed at 460 nm.

The quantum yield of photosynthesis decreased very dramatically with increasing wavelength beyond 685 nm, even though chlorophylls still absorb light at these wavelengths. This fact, the so-called *red drop* in photosynthesis, could not be explained at that time. However, Emerson and co-workers later showed that the amount of photosynthesis in far red light (wavelengths greater than 685 nm) could be increased considerably by a supplementary beam of red light (about 650 nm) (see Fig. 4.8). In fact, the total amount of photosynthesis carried out in the presence of a mixture of far red light and red light was greater than the sum of the amounts of photosynthesis carried out in separate experiments with the individual beams of light. This increase of the photosynthetic efficiency of far red light in the presence of a supplementary beam of lower wavelength is known as the *Emerson enhancement effect*. Experimental results of this effect are shown for three algae with different accessory pigments in Fig. 4.9. What is

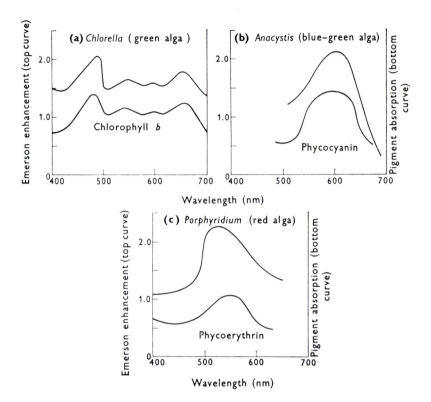

Fig. 4.9 Action spectrum of the Emerson effect in different algae (upper curve in each case) correlated with the absorption of the accessory pigments of the same algae (lower curve). (a) *Chlorella* containing chlorophyll *b*; (b) *Anacystis* containing phycocyanin; (c) *Porphyridium* containing phycoerythrin.

actually measured is O_2 evolution. From studies on the action spectrum of photosynthesis and of chlorophyll fluorescence, Emerson and Rabinowitch concluded that photosynthesis is enhanced by energy absorption by accessory pigments and transfer of this absorbed energy to chlorophyll *a* by the process of 'inductive resonance'. To explain the Emerson enhancement effect they put forward the hypothesis that at least two photochemical acts are involved in photosynthesis and are preferentially sensitized by different pigments, and that each electron must be photoactivated *twice* on its path from the primary donor, water, to the ultimate acceptor, CO_2. They also postulated the existence of two types of chlorophyll *a* in the cell, one associated with a reductant and another associated with an oxidant, and one of these is closer to the accessory pigment than the other – these are now termed photosystem I and photosystem II respectively.

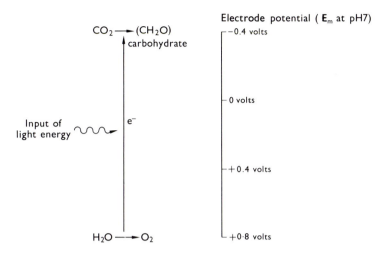

Fig. 4.10   Redox potentials of the overall reaction of photosynthesis.

Important subsequent work by Myers and French at Austin, Texas, in 1960 showed that photosystem I and photosystem II need not be excited simultaneously to get optimum photosynthesis, but that they may be illuminated alternately with a short dark period of a few seconds between them. This indicated that the two-light reactions could store their photochemical products for a short time before reacting with the electron transfer chain.

Experimental evidence and theoretical postulates in support of the two-light reaction hypothesis followed. The difference in electrode potential (ΔE) between the reactants and the final products in the photosynthetic reaction is 1.25 V (ΔE of CO_2–glucose couple $= -0.43$ V; ΔE of H_2O–O_2 couple $= +0.82$ V) (Fig. 4.10).

Photosynthetic reduction of CO_2 can be summarized by the equations:

$$2H_2O \xrightarrow{\text{light}} O_2 + 4H^+ + 4e^-$$
$$CO_2 + 4H^+ + 4e^- \longrightarrow (CH_2O) + H_2O$$

Thus 4 electrons are required to be transferred from water, through a redox span of 1.25 V, to reduce 1 molecule of CO_2. The energy required for the reduction of 1 mole of CO_2 is therefore $4 \times 1.25 \times 9.64 \times 10^4 = 48.2 \times 10^4$ J ($1\,eV = 9.64 \times 10^4$ J). Theoretically, this energy requirement can be satisfied by the capture of 4 mole quanta (1 photon per electron) of photosynthetically active red light, say 700 nm, which have an energy content of $4 \times 17.1 \times 10^4 = 68.4 \times 10^4$ J. However, due to thermodynamic losses

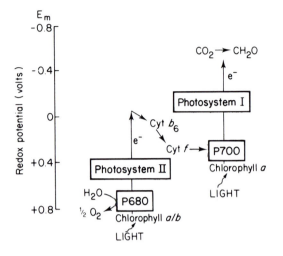

Fig. 4.11   The two-light reactions of the 'Z' scheme of photosynthesis. (Concept of Hill and Bendall.)

during energy conversion, only a fraction of the absorbed photon energy is converted to chemical free energy – in chloroplast photosynthesis this fraction seldom exceeds 0.36. Thus the maximum chemical free energy available for photosynthesis from a mole quantum of 700 nm light is $17.1 \times 10^4 \times 0.36 = 6.156 \times 10^4$ J. The minimum number of 700 nm quanta required to reduce a molecule of CO_2 is therefore $48.2 \times 10^4/6.156 \times 10^4 = 8$. If we accept this value of 8 as the quantum requirement for photosynthesis, then, since 8 quanta are consumed in the transport of 4 electrons (for 1 O_2 evolved), and since each quantum can activate only 1 electron at a time, it is logical to infer that each of the 4 electrons has to be activated by two separate light reactions.

In 1960, Hill and Bendall put forward the idea that the two light reactions should be in series (and not in parallel), with cytochromes b_6 and f acting as electron carriers in the non-photochemical reaction which connects the two photosystems. This is shown in Fig. 4.11.

Meanwhile, in Holland, Duysens, who refined the use of fluorescence measurements in the study of photosynthesis, observed that quanta absorbed in the red alga, *Porphyridium cruentum* at the Chl *a* maximum (430 nm) were less active in exciting Chl *a* fluorescence than quanta absorbed by the accessory pigment phycoerythrin at its maximum of 550 nm. Similarly, quanta at 670 nm, where only Chl *a* shows strong absorption, had the same

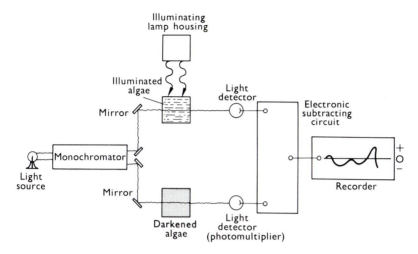

Fig. 4.12   Diagram of a difference spectrophotometer to measure the absorption of pigments.

low yield of fluorescence emission. Duysens concluded that two forms of chlorophyll *a* are present in the alga: a fluorescent form to which quanta absorbed by phycoerythrin were transferred (now known as PSII Chl *a*), and a relatively non-fluorescent form which has a strong absorption spectrum but very weak fluorescence excitation (PSI Chl *a*). The most conclusive evidence for two separate photosystems came from a series of difference spectroscopy studies initiated by Duysens, by Kok at Baltimore, and elaborated further by Witt in Germany. In this type of measurement, the change in absorbances of various constituents of a cell suspension is studied by illumination with monochromatic light of varying wavelengths. A schematic diagram of a difference spectrophotometer is shown in Fig. 4.12. By following the qualitative and quantitative changes in the difference spectrum of individual photosynthetic reaction components, e.g. cytochromes, the role of some of these components in the electron transfer pathway can be inferred. For example, Duysens illuminated a suspension of the red alga *Porphyridium* in the presence of DCMU (a synthetic weed killer which inhibits oxygen evolution) and found an accumulation of cytochrome *f* in the oxidized form. When the experiment was repeated with unpoisoned alga, there was no change in the cytochrome *f* spectrum. Duysens concluded that cytochrome *f* was an intermediate in the two light reactions, causing O_2 evolution (see also Chapter 5).

4.5   Reaction centres and primary electron acceptors

By illuminating suspensions of algae with brief flashes of light and following their optical spectra, Kok was able to identify and characterize a special type of chlorophyll *a* which he called P700. P700 is a trace constituent of chloroplasts with an absorption peak around 700 nm. It is reversibly bleached in light; the bleaching corresponds to an oxidation of P700 to $P700^+$. Subsequent investigations have shown that P700 is the reaction centre 'trap' pigment in which all incident light energy of wavelength greater than about 680 nm is captured and utilized for primary photochemical reactions, i.e. P700 is the primary electron donor of PSI. P700 is a special type of chloroplhyll *a*; evidence available at present is not conclusive as to whether it is a monomer or a dimeric chlorophyll molecule. The most stable electron acceptor from P700, ferredoxin, was purified from leaves and algae in the early 1960s. Since then a number of electron mediator species functioning between P700 and ferredoxin have been detected, mainly by measuring optical absorption changes and EPR signals which result from flash excitation of chloroplasts or purified PSI particles at very low temperatures (liquid N_2 and He). These are discussed in §8.9 and in Table 3.2.

Photosystem II has a higher content of chlorophyll *b* than PSI and it is strongly fluorescent. The reaction centre chlorophyll of PSII is designated as P680. Witt and co-workers followed chlorophyll absorption changes and rates of O_2 evolution in flashing light experiments in normal and DCMU-inhibited chloroplasts. They found a correlation between the O_2 evolution rate and the photo-induced absorption change of a chloroplast component with a peak around 682 nm. Further investigations have shown that P680 is also a specialized type of chlorophyll *a*. The primary electron acceptor from $P680^+$ in PSII has been identified as a pheophytin. From pheophytin the electron migrates, via two quinones Q_A and Q_B, to a plastoquinone pool which serves as an electron reservoir between the two photosystems. More detailed accounts of PSI and PSII are given in §8.9 and 8.8.

4.6   Experimental separation of the two photosystems

Techniques are now available to separate physically PSI and PSII from chloroplasts. Chloroplasts can be fragmented by the addition of detergents (digitonin, sodium dodecyl sulphate, Triton X-100, octylglucoside, sulfa-

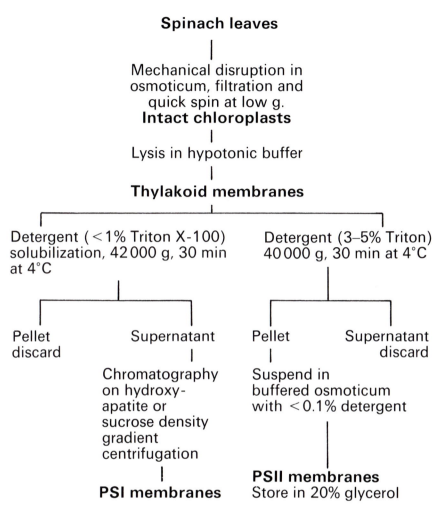

Fig. 4.13 A simplified scheme for the preparation of PSI- or PSII-enriched membranes from leaf tissue. The concentration of detergent (Triton X-100, digitonin, lauryl maltoside, octyl glucoside) used for solubilization of thylakoids depends on the chlorophyll content of the thylakoid suspension.

betain, etc.), by sonic vibration or by mechanical extrusion under high pressure (about 800 atmospheres) through a French or Yeda press. The fragments are then separated by a combination of differential and density gradient centrifugation and chromatography techniques (Fig. 4.13). Some of the preparations thus obtained had a very high chlorophyll *a* to chlorophyll *b* ratio and a high P700 content, suggesting they are highly enriched in PSI constituents. Particles enriched in PSI or PSII constituents have also been isolated from blue-green algae. Plant and algal mutants lacking either PSI or PSII have been produced by genetic manipulation and this has helped in the isolation of pure PS particles.

A PSII reaction centre core complex consisting of four chlorophyll and two pheophytin molecules and six polypeptides (47-kDa, 43-kDa; two 32-kDa (D_1 and D_2), and the 9 and 4-kDa subunits of cytochrome b559) was isolated from spinach grana thylakoids in 1987 by Nanba and Satoh in Japan. This spinach PSII complex, like the purple bacterial reaction centre complex (§7.3), is able to catalyze light-induced primary charge separation and electron transport, although it could not catalyze O_2 evolution.

By treatment with detergents, the proteins tightly bound to the thylakoid membranes can be brought into solution. The soluble components can then be resolved into fractions of varying molecular weights by electrophoresis on gels, e.g. polyacrylamide (PAGE), and located as specific bands in the gel by staining with dyes. Also it is possible to select mutants of plants and algae lacking specific chloroplast pigments or proteins and often deficient in some key photosynthetic activity or spectral characteristic. By comparing the gel bands from chloroplasts of such mutant and wild type plants, it is possible to assign a specific function to an individual protein or pigment complex identified in the gel. Some of the thylakoid components separated and tentatively identified by the above techniques are: a P700–chlorophyll a-protein complex associated with the PSI reaction centre, a light-harvesting chlorophyll 'a–b' protein complex associated with the PSII reaction centre, membrane-bound iron–sulphur proteins, cytochrome f, ATPase-coupling factor components, etc. (Fig. 4.14 and Plate I).

4.7 Inside-out and right-side-out chloroplast vesicles

The technique of aqueous-polymer two-phase partition for the subfractionation of chloroplasts has been introduced by Albertsson's group in Sweden. In this method, granal chloroplasts (see Fig. 3.3) suspended in a buffer containing 150 mM NaCl or 2 mM $MgCl_2$ (cationic medium to maintain grana stacking) are disintegrated by passing through a Yeda press at very high pressure of nitrogen. The Yeda press extrusion causes rupture of the peripheral granal membranes but the granal membranes stay appressed in the interior. After rupture, resealing of the appressed pairs of membranes from adjacent thylakoids results in the formation of vesicles with 'inverted'

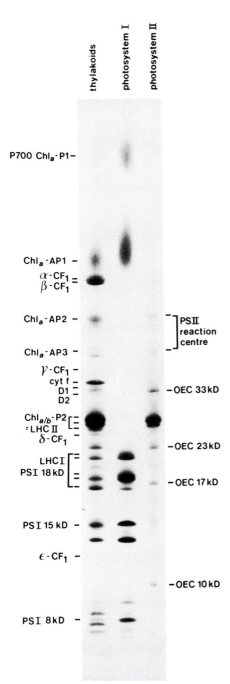

Fig. 4.14 Polypeptide patterns, obtained by polyacrylamide gel electrophoresis of barley thylakoids, PSI particles, and PSII particles. AP, apoprotein; CF_1, coupling factor extrinsic to the membrane; OEC, oxygen-evolving complex. See also Plate I. (Courtesy: D. Simpson and D. von Wettstein, Carlsberg Laboratory, Copenhagen.)

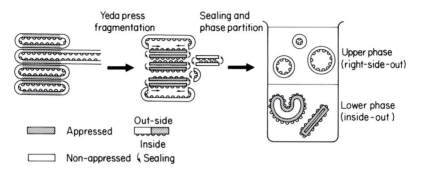

Fig. 4.15 Schematic illustration of the formation of right-side-out and inside-out membrane vesicles from thylakoids.

sidedness. These vesicles are called *inside-out vesicles* to distinguish them from normal thylakoid vesicles which are termed *right-side-out* (Fig. 4.15).

The inside-out and right-side-out vesicles differ in their surface charges and are separated by aqueous polymer phase partition, for example in a 5.7% dextran ($M_r \sim 500\,000$), 5.7% polyethylene glycol ($M_r \sim 3500$) and 88.6% water system. The inside-out vesicles settle in the dextran-rich lower phase, while the normal right-side-out vesicles accumulate in the polyethylene glycol-rich upper phase.

The vesicles thus fractionated have proved valuable in studies related to the topography and function of various components in the thylakoid membrane. Inside-out vesicles are rich in photosystem II components, suggesting that this photosystem is located in the granal membranes whereas photosystem I is concentrated more in the agranal membranes.

4.8 Photosynthetic oxygen evolution

The water-splitting (oxygen-evolving) reaction

$$2H_2O \rightarrow O_2 + 4H^+ + 4e^-$$

is carried out on the internal surface of the thylakoid membrane on the oxidizing side of the PSII. Each quantum captured by the PSII reaction centre generates one P680*. The P680* instantly transfers an electron to pheophytin (Pheo). P680$^+$ is reduced to P680 by the removal of an electron from Y_Z, the primary donor to PSII; Y_Z is a Tyr residue in the D_1 protein. The oxidized species Y_Z^+ immediately accepts an electron from the oxygen-evolving complex (designated M), which ultimately removes four

PSII Reaction Centre

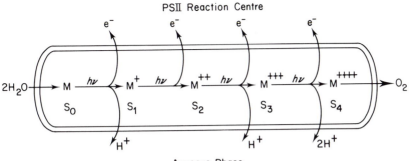

Fig. 4.16 Schematic illustration of the formation of S states and ejection of protons during photosynthetic water-oxidation resulting in O_2 evolution.

electrons from water (one at each step), so releasing O_2. The events can be schematically represented by the equations:

$$P680 \xrightarrow{h\nu} P680^*$$

(i) $\quad P680^* + Pheo \longrightarrow Pheo^- + P680^+$

(ii) $\quad P680^+ + Y_Z \longrightarrow P680 + Y_Z^+$

(iii) $\quad Y_Z^+ + M \longrightarrow Y_Z + M^+$

(iv) $\quad M^{4+} + 2H_2O \longrightarrow M + O_2 + 4H^+$

Notice that four positive charges need to be accumulated on M before an oxygen molecule can be released.

Kok, and Joliot (in Paris), measured oxygen evolution from dark-adapted chloroplasts and algae illuminated by a sequence of saturating light flashes. They observed that the maximum yield of O_2 per flash was obtained after the third flash and subsequently after every fourth flash, i.e. O_2 peaks occurred at flashes 3, 7, 11, etc. To explain these periodic O_2 peaks, Kok and Joliot independently proposed the '*S state hypothesis*'. The O_2-evolving complex M can cycle through five different oxidation states, viz. S_0, S_1, S_2, S_3 and S_4, each state differing from the preceding one by the loss of a single electron (Fig. 4.16). When four electrons are lost by M, i.e. when M has reached the S_4 state, it is able to react with water, producing O_2 and returning to the uncharged state, S_0. Dark adaptation of chloroplasts synchronizes most of the O_2-evolving centres to the S_1 state – hence maximum O_2 evolution occurs after the third flash at the start of the flash regime. During water oxidation, protons are released into the interior of the thylakoid vesicles, the sequence of proton release probably occurring as shown in Fig. 4.16.

The structure of the O_2-evolving complex M and the exact mechanism of water oxidation are still not clearly understood. Studies using inhibitors of O_2 evolution, spectroscopic data, and reconstitution experiments suggest that M is a mangano–protein complex containing four atoms of Mn, two of which are essential for water oxidation. Reagents such as NH_2OH and NH_2NH_2 at low concentrations reduce the S_1 centres and inhibit O_2 evolution. Heat treatment of chloroplasts (55°C for 5 min) and washing in 1 M Tris, 1 M divalent salts, or in alkaline buffer (pH 8) dissociate the proteins of the complex M, release Mn, and inhibit O_2 evolution. Depletion of chloroplasts of chloride ions, or in some instances of calcium ions, also inhibits O_2 evolution – these ions may play a physiological role in maintaining the integrity of M.

More details on water oxidation are given in §8.7.

5

Photosynthetic electron transport and phosphorylation

In Chapter 2 we elucidated the idea of photosynthesis involving both light and dark phases in the fixation of CO_2. The recognition and experimental demonstration of these two phases was an important step towards the modern understanding of the CO_2-fixation process. This was not possible until Arnon, Allen and Whatley in 1954 were able to isolate chloroplasts from spinach leaves which were capable of carrying out complete photosynthesis, i.e. fixing CO_2 to the level of carbohydrate. They were able physically to separate the light and dark phases and to show the light-dependent formation of ATP and $NADPH_2$, which then acted as the energy sources for the subsequent dark fixation of CO_2. This is summarized in the familiar diagram below (Fig. 5.1).

The light phase, which occurs subsequent to the initial light reactions discussed in the previous chapter, involves biochemical reactions with life times of 10^{-2} to 10^{-5} s. The initial light reactions have of course much shorter lifetimes – down to 10^{-15} s. The biochemical events of the light phase result in (*i*) the production of the strong reducing agent, $NADPH_2$, (*ii*) the accompanying evolution of O_2 as a by-product of the splitting of

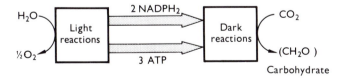

Fig. 5.1 Major products of the light and dark reactions of photosynthesis.

H_2O, and (iii) the formation of ATP which is coupled to the flow of electrons from H_2O to NADP.

In this chapter we shall discuss how these reactions are thought to occur, what evidence there is for making such assumptions, and what compounds are involved in the sequence of electron transport reactions.

5.1   Reduction and oxidation of electron carriers

The production of $NADPH_2$, ATP and O_2 in the lamellae of the chloroplast involves the transfer of electrons through a chain of electron carriers. This electron transfer requires that each of the carriers in turn becomes reduced and oxidized in order that the energy in the electron can be passed along the chain. Reduction simply means the adding of an electron, while oxidation implies the removal of an electron from a compound. Whenever an electron is exchanged between two compounds, one is oxidized and the other reduced. Almost every such exchange is accompanied by the release or absorption of energy. It makes no difference whether we think of the energy as arising out of the pull exerted on the electron by 'oxidizing power' or the push exerted by 'reducing power'.

Often, though not invariably, an electron travels in company with a proton, i.e. as part of a hydrogen atom. In that case oxidation means removing hydrogen and reduction means adding hydrogen. Thus NADP is reduced to $NADPH_2$ and CO_2 to carbohydrate, by the addition of hydrogen atoms.

The oxidation–reduction ('redox') potentials of biological electron carriers are expressed on a voltage scale at biological pHs, which indicates that the $H_2O \rightarrow O_2$ couple is very oxidizing, with a positive midpoint potential of $+0.82$ V, while the $H^+ \rightarrow H_2$ (gas) couple is very reducing, with a negative potential of -0.42 V. Most biological electron transfer reactions occur between these two extremes (see Table 5.1). We shall see later that, in fact, the process of photosynthetic electron transport takes place at between $+0.82$ V and -0.42 V. In order to transfer electrons between these extremes of redox potential, light energy is required.

5.2   Two types of photosynthetic phosphorylation

Photosynthetic phosphorylation is the production of ATP in the chloroplast or in other membranes by light-activated reactions. It can take place in

Table 5.1. *Midpoint redox potentials [E_m] of some chloroplast components and reactions*

Chloroplast component or reaction	E_m (volts)
P680 of PSII	+0.9 (or more +ve)
H_2O/O_2	+0.82
P700 of PSI	+0.48
Cytochrome *f*	+0.35
Plastocyanin	+0.38
[Fe–S]$_R$	+0.29
Plastoquinone	0.0
Cytochrome b_6	−0.05 and −0.17
NADPH$_2$	−0.34
H^+/H_2	−0.42
Ferredoxin	−0.42
CO_2/CH_2O	−0.43
F_A of PSI	−0.55
F_B of PSI	−0.59
F_X of PSI	−0.73

the chloroplasts via two processes: non-cyclic and cyclic. In non-cyclic photophosphorylation, ATP is generated in an 'open' electron transfer system together with the evolution of O_2 from H_2O and the formation of NADPH$_2$ from NADP. In cyclic photophosphorylation the electrons cycle in a 'closed' system through the phosphorylation sites and ATP is the only product formed. These two systems are shown in Fig. 5.2. The role of cyclic photophosphorylation *in vivo* is still being debated.

5.3 Non-cyclic electron transport and phosphorylation

This is the light-requiring process in which electrons are removed from H_2O resulting in the evolution of O_2 as a by-product (Hill reaction, Chapter 2), and the transfer of these electrons via a number of carriers to produce a strong, negative reducing potential with the subsequent formation of NADPH$_2$, a reducing agent with a potential of −0.34 V.

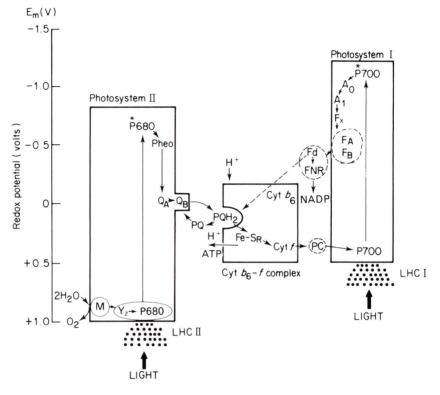

Fig. 5.2 Electron transport scheme in chloroplasts. Rectangular boxes represent membrane-bound complexes. LHC I and LHC II, light-harvesting complexes of PSI and PSII respectively; other symbols as in Table 3.2.

This is simply expressed as:

$$NADP + H_2O \xrightarrow[\text{chloroplasts}]{\text{light}} NADPH_2 + \tfrac{1}{2}O_2$$

The carriers which have been identified include chlorophyll a, pheophytin, quinones, cytochromes b and f, Fe–S centres, plastocyanin, ferredoxin and FNR, a flavoprotein enzyme.

ATP formation accompanies the transfer of electrons and protons via the quinones and involves side reactions which are obligatorily coupled to the electron transfer process. Thus $NADPH_2$ and ATP formation occur in the process of non-cyclic electron transport where electrons are removed from

H_2O and donated to NADP with the coupled formation of ATP. This overall reaction can be expressed as:

$$2NADP + 2H_2O + 2ADP + 2Pi \xrightarrow[\text{chloroplasts}]{\text{light (4e}^-)} 2NADPH_2 + 2ATP + O_2$$

This equation implies that each H_2O is split in the chloroplast membrane under the influence of light to give off $\frac{1}{2}O_2$ molecule (an atom of oxygen), and that the two electrons so freed are then transferred to NADP, along with H^+s (protons), to produce the strong reducing agent, $NADPH_2$. Two molecules of ATP can be simultaneously formed from two ADP and two Pi (inorganic phosphate) so that energy is stored in the form of this high energy compound. The precise number of ATP molecules formed is unclear, although it is generally thought to be two per O_2 evolved.

The $NADPH_2$ and ATP are the 'assimilatory power' required to reduce CO_2 to carbohydrate in the dark phase, which will be discussed in the next chapter. This 'assimilatory power' represents the initial products of the conversion of light energy into chemical energy.

A diagrammatic representation of the electron flow pattern in non-cyclic phosphorylation is given in Fig. 5.2. This formulation is derived from the Z scheme of Hill and Bendall. The scale on the left shows very clearly the potential of all the electron carriers in the chain and implies that the sequence of electron flow depends to a large extent on their potential. This is a very neat view of electron transport and all the evidence to date indicates that it is most probably correct. The components of the non-cyclic electron transport pathway are organized into three complexes which span the chloroplast membrane. These complexes are the PSII complex, cyt $b\,f$ complex and the PSI complex, and they can be separated from the membrane by detergent treatment. Plastoquinone, plastocyanin and ferre-doxin are the mobile carriers which shuttle electrons across the complexes. An electron is transferred from H_2O to NADP in about 20 ms.

What is immediately apparent is that two different light reactions are required to raise the electrons from the level of H_2O ($+0.82$ V) to the level of $NADPH_2$ (-0.34 V): these are designated photosystems II and I. Each of the systems has a different type of chlorophyll as the main light-absorbing pigment. Photosystem I has a predominance of chlorophyll a with an absorption maximum of 680 nm, (bluish-green in colour), while photosystem II also contains the closely related chlorophyll b (see Chapter 3) which has its main absorption peak at 650 nm (yellowish-green in colour).

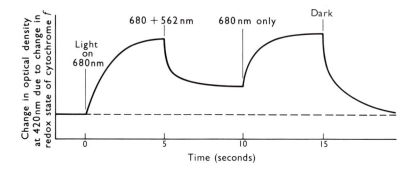

Fig. 5.3 Oxidation and reduction of cytochrome *f* in the red alga *Porphyridium*. Increase in OD420 is due to oxidation and decrease in OD420 is due to reduction of cytochrome *f*. Light of 680 nm is absorbed by photosystem I (chlorophyll *a*) and at 562 nm is absorbed by photosystem II (phycoerythrin).

The first evidence for the possible involvement of two light reactions in photosynthesis came from the work of Emerson and co-workers from 1943 onwards (see also Chapters 2 and 4). They showed that the reduction of one molecule of CO_2 to carbohydrate required light of two different wavelengths – the quantum yield was very low if one light which was *only* absorbed by chlorophyll *a*, e.g. light of wavelength greater than 680 nm, was used. The wavelength of the second light required to give efficient photosynthesis was found to correspond to that of chlorophyll *b* in higher plants and green algae and to the other accessory pigments in red and blue-green algae, e.g. phycoerythrin and phycocyanin.

Further evidence for the requirement of light of two different wavelengths in non-cyclic electron flow from H_2O to NADP came from experiments measuring the changes in oxidation and reduction state of the cytochromes in algal chloroplasts by Duysens and Amesz. In Fig. 5.2 cyt *f* is an electron carrier intermediate between photosystems I and II. Using very sensitive spectrophotometric techniques (see §4.4) it is possible to measure the redox state of the cyt *f*, due to its specific absorption peaks at 422 and 550 nm, under illumination of light of different wavelengths being absorbed by the two photosystems. The experimental results are shown in Fig. 5.3. Cytochrome in the chloroplast is naturally in the reduced state in the dark. If the algae are illuminated with light of 680 nm wavelength, which is mainly absorbed by photosystem II., i.e. chlorophyll *a* in the red alga *Porphyridium*, electrons are removed from cyt *f* and donated to ferredoxin and thence to NADP – thus the cyt *f* becomes oxidized. Then if light at 562 nm (absorbed by photosystem II, phycoerythrin in *Porphyridium*, but also to some extent by photosystem I) is applied to the chloroplast, the cyt f becomes reduced since it accepts electrons from photosystem II. Note that complete reduction is not achieved since photosystem I is still functioning to a limited extent in removing the electrons from the cyt f. In the dark the cyt f returns to its normal reduced state seen at the start of the experiment. This type of experiment was initiated by Duysens in 1961 and has proved of great value in localizing the site of electron carriers in the chain.

Many different compounds have been isolated from chloroplast membranes, but the electron carriers shown in the scheme for non-cyclic electron transport (Fig. 5.2) are those for which a role has been assigned so far. These compounds have been isolated and characterized chemically and spectroscopically.

Experiments have been devised which enable one to remove a specific carrier from the chloroplasts, e.g. by washing with water or low concentrations of detergents, organic solvent extraction or by mild sonication; a certain electron transfer step is thus inhibited and then the re-addition of the extracted compound in its purified form restores the electron-carrying ability of the chloroplast membranes. This type of experiment has been successfully applied to define the location of plastoquinone, plastocyanin, ferredoxin, and FNR, which is the reductase enzyme (a flavoprotein) acting between ferredoxin and NADP.

Elegant experiments by Levine (in Harvard), von Wettstein (Copenhagen), Bishop (Oregon), Ohad (Israel) and others using genetic mutants of barley and the green algae *Chlamydomonas* and *Scenedesmus* have also helped to consolidate the electron flow sequence of Fig. 5.2. Specific mutants of the algae have been obtained which are deficient in certain parts of the electron chain, e.g. blocking electron flow between PQ and cyt f or between plastocyanin and P700. With a knowledge of the exact site of the blockage, experiments can be designed in which electrons are donated to or withdrawn from various parts of the chain. This can be quite easily accomplished by using different types of redox dyes and was also used successfully in the spectrophotometric experiments mentioned earlier. These genetic and biochemical experiments are similar in principle to those used earlier with the mould *Neurospora* and the bacterium *Escherichia coli*, in studies on electron flow and energy transduction across the membranes.

Since 1965, low-temperature (liquid N_2, 77 K and He, 4 K) EPR has been used as a valuable probe in the identification of paramagnetic species as components of the chloroplast electron transport chain. This technique, coupled with measurements of redox potentials under anaerobic conditions, has enabled investigators to assign locations for components such as pheophytin, $[Fe-S]_R$ and F_X, F_A, F_B and A_1 of PSI. The technique is particularly useful to characterize membrane-bound iron–sulphur proteins.

The use of chemical inhibitors of specific biochemical reactions is a classical and fruitful approach to understanding biochemical mechanisms. Non-cyclic photophosphorylation is no exception to the advantageous use of specific inhibitors. A number of different inhibitors have been found which block specific parts of the chain, e.g. DCMU [− (3,4-dichlorophenyl) − 1, 1-dimethyl urea], a herbicide which blocks oxygen evolution by binding to the Q_B protein of the PSII complex; antimycin A, an antibiotic, which prevents the reduction of cytochrome f; and KCN, which inhibits plastocyanin (see §5.7).

5.4   ATP synthesis in chloroplasts: the chemiosmotic hypothesis

We will now discuss the ATP formation which is coupled to the non-cyclic electron flow. In 1958 Arnon, Whatley and Allen demonstrated the obligatory coupling of ATP formation to the reduction of NADP and showed that the rate of electron flow to NADP was dependent on the presence of ADP and Pi (required to form ATP). How many ATP molecules are formed per $NADPH_2$ produced is an important question because of the amount of ATP needed to fix CO_2 in the dark phase – two $NADPH_2$s and three ATPs are required per CO_2 molecule reduced to the level of carbohydrate in C_3 plants. This necessitates an $ATP:2e^-$, ratio of at least 1.5.

The chemiosmotic hypothesis

Mitchell (working in Edinburgh) in 1961 proposed a *chemiosmotic hypothesis* to account for ATP formation accompanying electron transport in mitochondria and chloroplasts; for this he was awarded the Nobel Prize in 1978. The chemiosmotic theory envisages that (*1*) intact membranes are almost impermeable to the passive flow of protons, (*2*) the electron donors, and acceptors of electron and proton are arranged vectorially in the membrane, and (*3*) during photosynthetic electron transport, protons are translocated from the external medium (stroma) into the osmotic space of the intrathylakoid membrane. Thus on illumination of chloroplasts, for each quantum transferred to the reaction centre (P680 or P700), an electron is excited and transported through the membrane; simultaneously, a proton is taken up from the outside to the inside of the membrane. As a result of the movement of protons (and electrons) across the membrane, the membranes become energized and generate a transmembrane electrochemical potential referred to as the proton motive force (pmf). The magnitude of this proton motive force is expressed by the relation:

$$\text{pmf (volts)} = 2.303 \frac{RT}{F} \Delta pH + \Delta \psi$$

where R is the gas constant (8.314 J mol^{-1}/$^{\circ}$K), F is the Faraday constant (96.5 kJ mol^{-1} V^{-1}), T is temperature in the absolute scale ($^{\circ}$K), ΔpH is the transmembrane pH gradient, and $\Delta \psi$ the membrane electrical potential in volts. The energy available for ATP synthesis is equal to the product of the pmf and the number of protons consumed per molecule of ATP synthesized. According to the chemiosmotic hypothesis, synthesis of ATP is promoted by the back-flow of protons via the enzyme ATP-synthase (also called the coupling factor), located across the membrane. The energy released by the flow of protons is utilized for the synthesis of ATP, an endergonic reaction:

$$\text{ADP} + \text{Pi} \rightarrow \text{ATP} \quad \Delta G = 30 \text{ kJ (7.3 kcal)}$$

where ΔG is the change in Gibb's free energy (G).

There are two sites in the thylakoid membrane where proton uptake accompanies electron transfer. The secondary quinone Q_B of PSII after double reduction to Q_B^{2-} picks up two protons from the stromal side of the membrane and moves to the cyt bf complex as plastoquinol. The second proton uptake reaction occurs on the periphery of the PSI complex where NADP is reduced to $NADPH_2$ (by the addition of H^+ and e^-), a reaction catalyzed by the enzyme FNR, with electrons donated by reduced Fd. Protons are released into the thylakoid lumen during electron transport by (a) water oxidation at PSII, and (b) plastoquinol oxidation via the cyt bf complex (Fig. 5.2 and §8.10).

Data from several types of experiments support Mitchell's hypothesis. Firstly, if chloroplasts are illuminated under non-phosphorylating conditions in a medium depleted of ADP, they catalyze an electron transport-dependent proton uptake, thereby generating a transmembrane proton and electron gradient, both of which can be easily measured. The increase in intrathylakoid pH thus generated is of the order of 3 to 4 pH units; the thylakoid membranes can withstand a proton concentration gradient of $10\,000:1$. Secondly, as Neumann and Jagendorf at Johns Hopkins Univer-

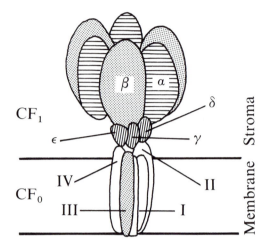

Fig. 5.4   A schematic representation of the chloroplast ATP synthase (side view) showing the organization of the subunits within the CF_1 and CF_0 complexes. (Redrawn from Ort and Oxborough, 1992.)

sity demonstrated in 1964, if broken chloroplasts are pre-incubated in the dark in an appropriate buffer of low pH (pH = 4) and then rapidly injected into an alkaline buffer (pH = 8) containing ADP and Pi, the synthesis of ATP occurs even in the dark due to the establishment of a pH (proton) gradient. Although other theories to explain the phenomenon of ATP synthesis (photophosphorylation) have been put forward, the chemiosmotic theory has gained wide acceptance among biochemists.

ATPase: coupling factors

The synthesis of ATP from ADP, Pi and protons is catalyzed by the protein complex, ATPase or ATP-synthase, also known as the coupling factor (CF) since it couples electron (and proton) flow to phosphorylation. In chloroplasts ATPase is located mainly in the non-granal (stromal) membrane (see Plate I) and is composed of two distinctive components: factor CF_1, a hydrophilic protein projecting towards the stroma, and CF_0, a hydrophobic protein which spans across the membrane. CF_1, which was detected in chloroplasts by Avron (in Israel) in 1963, can be easily released from the membrane into aqueous buffers, has ATP-hydrolase activity (after suitable treatment) and has a molecular weight of 400 kD. It is composed of five subunits named α, β, γ, δ, and ϵ, with α and β subunits existing as trimers (Fig. 5.4).

The membrane-bound component of the coupling factor, CF_0, contains four different polypeptides referred to as subunits I to IV. The functions of the various subunits of the complex have been deduced from chemical modification, reconstitution, and kinetic studies. Two $\alpha\beta$ pairs of CF_1 are thought to contain one catalytic and one regulatory site. The third $\alpha\beta$ pair, which is in closer contact with the δ, γ, ϵ subunits, binds two structural nucleotides (ATP/ADP) and has no catalytic site. CF_0 acts as a channel for pumping protons across the membrane to CF_1.

ATP formation begins just a few milliseconds after illumination of chloroplasts; the reverse reaction (ATP hydrolysis by the CF complex) is sluggish and requires a few seconds of illumination. Thus the coupling factor plays a dynamic regulatory role that restricts the hydrolysis of stromal ATP but at the same time promotes efficient ATP synthesis when energetic conditions allow. Experimental evidence indicates a regulatory mechanism involving three principal factors: the transmembrane electrochemical potential ($\Delta\psi$), the oxidation state of the γ subunit cystine bridge, and the relative binding constants of the ATP, ADP, and Pi to the ATP-synthase complex. The activation of ATP-synthase in the light requires reduced thioredoxin (§8.15) formed in PSI during photosynthetic electron transport; the reduced thioredoxin is thought to reduce a cystine disulphur bridge in the γ subunit of CF_1 to cysteines.

Uncouplers

Chemical reagents that stop ATP synthesis in membranes while allowing electron transport to proceed are known as *uncouplers*. These compounds allow the passive movement of protons across the membrane and restrict the formation of a proton gradient essential for phosphorylation. Chemicals such as carbonyl cyanide 3-chlorophenyl hydrazone (CCCP), and amines, make the membrane more permeable to protons. Carbonyl cyanide p-tri fluoromethoxyphenyl hydrazone (FCCP) and CCCP contain dissociable protons and are able to permeate the lipid bilayer membranes as protonated bases or conjugated bases; they are known as proton translocation uncouplers. Weak bases, e.g. methylamine or ammonia in the neutral state, can readily diffuse into the thylakoids and there combine with H^+ ions; the protonated bases $CH_3NH_3^+$ and NH_4^+ cannot diffuse out of the membrane easily. The natural antibiotics gramicidin and valinomycin are mobile ion carriers which can diffuse across the membrane and catalyze the transport of ions at a very fast rate through the membrane; such molecules

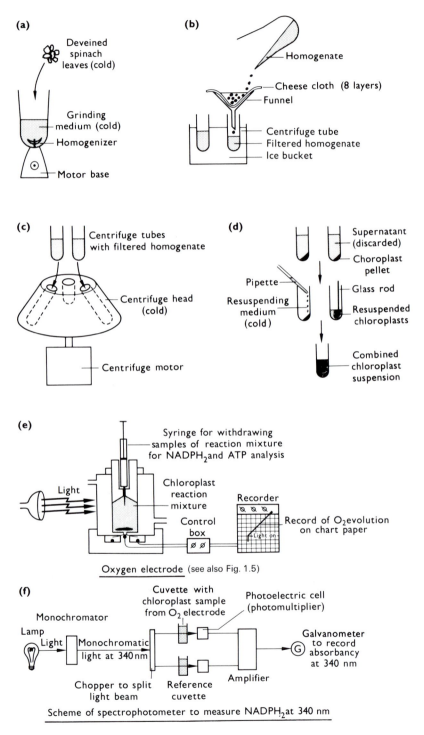

(a)

Deveined spinach leaves (cold)

Grinding medium (cold)

Homogenizer

Motor base

(b)

Homogenate

Cheese cloth (8 layers)

Funnel

Centrifuge tube

Filtered homogenate

Ice bucket

(c)

Centrifuge tubes with filtered homogenate

Centrifuge head (cold)

Centrifuge motor

(d)

Supernatant (discarded)

Choroplast pellet

Pipette

Resuspending medium (cold)

Glass rod

Resuspended chloroplasts

Combined chloroplast suspension

(e)

Syringe for withdrawing samples of reaction mixture for NADPH$_2$ and ATP analysis

Light

Chloroplast reaction mixture

Control box

Recorder

Record of O$_2$ evolution on chart paper

Light on

Oxygen electrode (see also Fig. 1.5)

(f)

Cuvette with chloroplast sample from O$_2$ electrode

Photoelectric cell (photomultiplier)

Monochromator

Lamp

Light

Monochromatic light at 340 nm

Galvanometer to record absorbancy at 340 nm

Chopper to split light beam

Reference cuvette

Amplifier

Scheme of spectrophotometer to measure NADPH$_2$ at 340 nm

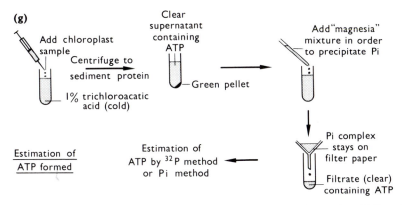

(g)

Fig. 5.5   Experiment to demonstrate photosynthetic O_2 evolution and $NADPH_2$ and ATP formation by isolated spinach chloroplasts.

are referred to as *ionophores*. The ionophore nigericin, when applied in conjunction with potassium ions, promotes a H^+/K^+ exchange across the membrane and thus dissipates the proton gradient. All these types of uncouplers have been used in studies to unravel the mechanism of photophosphorylation. ATP synthesis can also be blocked at the coupling site by inhibitors such as dicyclohexylcarbodiimide, triphenyltin, etc. and antibody to CF_1 which reacts directly with the coupling factors. These organic compounds are referred to as energy transfer inhibitors.

During the isolation and preparation of chloroplasts from the whole leaf, the ATP formation factors are easily destroyed, so that greater care must be taken in performing phosphorylation experiments than those in which only electron transport is measured. In Fig. 5.5 a sequence of diagrams shows how one isolates chloroplasts and measures the O_2 evolution, $NADPH_2$ formation and ATP synthesis associated with non-cyclic photophosphorylation.

5.5   Cyclic electron transport and phosphorylation

In this process, which requires light and chloroplasts, the only net product is ATP. The reaction was discovered in 1954 by Arnon, Allen and Whatley using isolated spinach chloroplasts and by Frenkel using chromatophores isolated from photosynthetic bacteria. It may be very simply represented by the following equation:

$$ADP + Pi \xrightarrow[\substack{\text{chloroplasts} \\ \text{(or chromatophores)}}]{\text{light}} ATP$$

Only a cyclic electron flow involving photosystem I is required in order to produce ATP. Figure 5.2 suggests how this may occur. Under the influence of an input of light, an electron is removed from P700 in its excited state and donated to Fe–S centres and subsequently to ferredoxin, which becomes reduced. The reduced ferredoxin then, instead of transferring its electron to NADP as in the case of non-cyclic electron flow, donates its electron to cytochrome b_6 and thence through the electron transport chain back to P700. Thus the electron undergoes a cyclic flow around PSI and cyt bf complex (without PSII involvement). The only measurable product is ATP, which is formed by a coupling mechanism, probably similar to that involved in non-cyclic photophosphorylation even though the electron carriers may not be identical. The number of molecules of ATP formed per electron transferred is so far undetermined because of the difficulty in measuring the number of electrons cycling around the chain in a given time – the number of ATPs formed in a given time is, on the other hand, relatively simple to measure.

We can thus see that ferredoxin may play a central role in photosynthesis. It can donate electrons in a non-cyclic system to NADP in order to produce the strong reducing power in the form of $NADPH_2$ needed for CO_2 reduction; or it can donate electrons back into the electron transfer chain in a cyclic system resulting only in the formation of ATP. This ATP can be used for CO_2 fixation or for other reactions which only require ATP as their energy source, e.g. protein synthesis and the conversion of glucose to starch, both of which occur in the chloroplast. The physiological controlling mechanism associated with ferredoxin and its very low reducing potential of -0.42 V (equivalent to that of H_2 gas) have stimulated much research into the physicochemical and biochemical properties of this unique protein. A model for the active centre of ferredoxin is given in Fig. 5.6a, and some reactions catalyzed by Fd are shown in Fig. 5.6b.

In the chloroplast, ferredoxin is the physiological carrier (or cofactor) involved in cyclic photophosphorylation. However, experimentally we can replace ferredoxin by vitamin K_3, FMN, or any of a number of dyes. The cyclic electron flow sequence may not be exactly the same as that for ferredoxin, but again the only net product is ATP.

Pseudocyclic electron transport   Under certain physiological conditions it is possible for reduced Fd (or other PSI electron acceptors) to react directly with molecular O_2, rather than with NADP, thereby forming H_2O_2. The formation of H_2O_2 with uptake of O_2 by illuminated spinach chloroplasts was first observed by Mehler in Chicago in 1951 and is called the *Mehler*

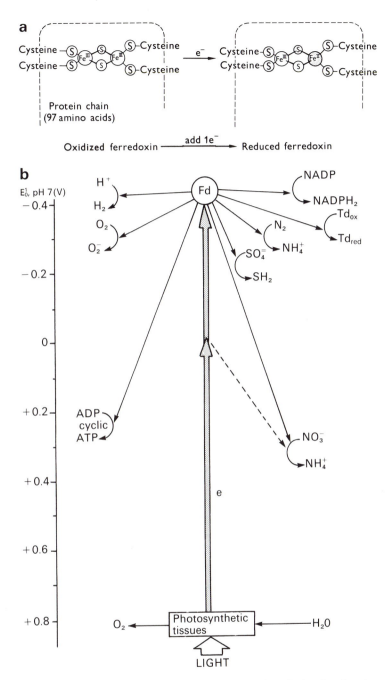

Fig. 5.6 (a) A model for the active centre of plant ferredoxins showing the oxidized and reduced forms of the protein. (b) Schematic illustration showing various metabolic reactions in chloroplasts (or cyanobacteria) catalyzed by ferredoxin. Td, thioredoxin, O_2^-, superoxide.

Table 5.2. *Carbon dioxide fixation in the dark and light by chloroplast systems, i.e. stroma (yellowish matrix) and grana (chlorophyll-containing, green membranes). (Trebst, Tsujimoto and Arnon (1958).* Nature, **182**, *351.)*

	$^{14}CO_2$ fixed (counts per minute)
Stroma (dark)	4 000
Stroma (dark) + grana (light)	96 000
Stroma (dark) + ATP	43 000
Stroma (dark) + NADPH$_2$ + ATP	97 000

Note: The equivalence of grana (light) and (NADPH$_2$ + ATP), i.e. assimilatory power.

reaction. The type of non-cyclic electron transport from H_2O to O_2 with formation of ATP (but no NADPH$_2$) is referred to as pseudocyclic electron transport. During pseudocyclic electron transport the initial product of oxygen reduction is the superoxide (O_2^-) radical, which is either hydrated to H_2O_2 or dismuted back to O_2 by the enzyme *superoxide dismutase* found in chloroplasts. Pseudocyclic electron transport provides extra ATP and removes O_2; these processes are important in the regulation of CO_2 assimilation and for the protection of chloroplasts under photoinhibitory conditions (§8.13).

In addition to O_2^- and H_2O_2, illuminated chloroplasts can generate singlet oxygen (O·) by transfer of photon energy from excited Chl to O_2. Carotenoids can decrease this problem by (a) reacting with the singlet oxygen, and (b) quenching the chlorophyll triplet states that lead to O· formation. Plant-produced antioxidants help protect against the harmful effects of oxygen species.

5.6   Structure – function relationships

The light phase of the overall CO_2 fixation to the level of carbohydrate has been shown to occur in the grana lamellae (or thylakoids) of the chloroplast, while the dark phase occurs in the stroma of the chloroplast. Arnon and co-workers demonstrated this in 1958 by physically separating the light and dark phase (see fig 2.3 and Table 5.2). Chloroplasts were illuminated in the absence of CO_2 and allowed to form large amounts of NADPH$_2$ and ATP, with the concomitant O_2 evolution from the non-cyclic electron flow. The chloroplasts were then broken and the stroma separated from the grana lamellae, which were discarded. Then in the dark, radioactive $^{14}CO_2$ was supplied and the enzymes in the stroma proceeded to assimilate the CO_2 to produce the same carbohydrates that whole chloroplasts and intact leaves synthesize.

These experiments very neatly showed that all the electron carriers and enzymes required for the light-induced $NADPH_2$ and ATP formation via cyclic and non-cyclic electron flow are associated with the chloroplast grana, and also that $NADPH_2$ and ATP are both required for the fixation of CO_2 by the enzymes which are found in the soluble stromal fraction.

5.7 Artificial electron donors, electron acceptors, and inhibitors

In Chapter 4 and in this chapter, various redox compounds which can donate electrons to or accept electrons from the photosynthetic electron transport chain have been mentioned. The use of these artificial electron mediators has been extremely useful in measuring the electron transport activity of various segments of the photosynthetic pathway in isolated chloroplasts. The activities usually assayed are oxygen exchange (evolution or uptake), optical absorption change, or induction and decay of chlorophyll fluorescence. The specificity of an artificial mediator is dictated by its redox potential and its accessibility to the electron transport chain. Photosystem II has a redox span of $+ 820$ mV (water oxidizing complex) to 0.0 mV (reduced plastoquinone) and it has the potential to oxidize any redox molecule with an E_m more negative than 820 mV and to reduce any molecule with an E_m more positive than 0.0 mV. Similarly, PSI can be expected to oxidize a mediator of E_m more negative than $+ 480$ mV (E_m of P700) and reduce a mediator of E_m more positive than $- 700$ mV (E_m of F_x).

The accessibility of an electron mediator to its specific site depends on its solubility in lipid or aqueous phase, i.e. whether the molecule is hydrophobic or hydrophilic – some molecules may be amphiphilic. Hydrophobic molecules are more accessible to PSII components which lie embedded in the thylakoid membrane. On the other hand, hydrophilic molecules are good electron acceptors from the reducing side of PSI which lies exposed on the outer side of the membrane.

Many electron mediators used to measure chloroplast electron transport are only stable either in the oxidized state or in the reduced state, and some of them inhibit electron transport when used in high concentrations. To

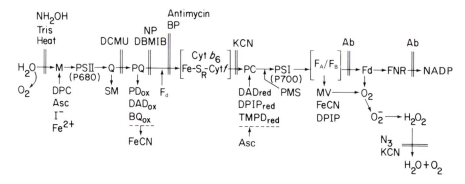

Fig. 5.7   Inhibitors of electron transport, and artificial electron donors and electron acceptors. Vertical double lines indicate probable site of inhibitor action; arrows towards the chain indicate site of electron donation; and arrows out of the chain indicate site of electron acceptance. Ab, antibody; Asc, ascorbate; BP, bathophenanthroline (iron-chelator); BQ, benzoquinone; DAD, diaminodurene (2, 3, 5, 6-tetramethyl-p-phenylene diamine); DBMIB, 2,5-dibromo-3-methyl-6-isopropyl-p-benzoquinone; DCMU, 3-(3,4-dichlorophenyl)-1, 1-dimethylurea; DPC, diphenylcarbazide; DPIP, dichlorophenol-indophenol; FeCN, ferricyanide; MV, methylviologen; NP, nitrophenol; O_2^-, superoxide; PD, phenylenediamine; PMS, phenazine methosulphate; SM, silicomolybdate; TMPD, N-tetramethyl-p-phenylene diamine.

overcome these drawbacks, such mediators are usually added to the reaction medium in catalytic amounts in conjunction with a stable, non-inhibitory oxidant or reductant as the case may be. For example, the PSII electron acceptors paraphenylene diamine (PD), diaminodurene (DAD), and benzoquinone (BQ) are maintained in the oxidized state in reaction mixtures by the addition of excess ferricyanide. Similarly, when used as electron donors to plastocyanin (and PSI), the compounds N-tetramethyl-p-phenylene diamine (TMPD), dichlorophenolindophenol (DPIP) or DAD are kept reduced by the presence of excess of ascorbate.

The accessibility of an electron donor or acceptor to the photosynthetic chain depends in addition on the structural integrity of the thylakoids in the chloroplast preparation. Thus, the hydrophilic compounds ferricyanide and $DPIP_{ox}$ are used as PSI acceptors in more intact chloroplasts but will accept electrons also from PSII in broken chloroplasts.

Electron transport can be blocked at specific sites by the addition of compounds which bind to one of the components of the chain, remove one of the components, or alter the fluidity of the membrane structure. Antibodies raised against the proteins in the electron transfer chain specifically block by binding to their antigens. The more commonly used electron acceptors, donors, and electron transport inhibitors are shown in Fig. 5.7.

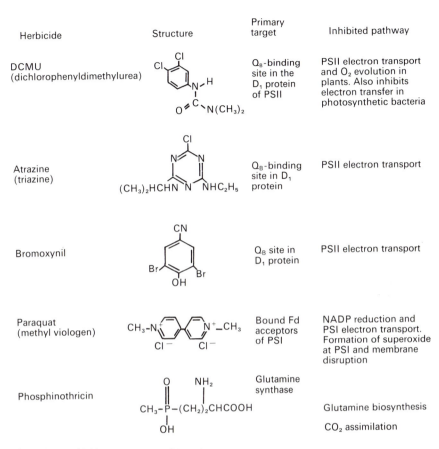

Herbicide	Structure	Primary target	Inhibited pathway
DCMU (dichlorophenyldimethylurea)		Q_B-binding site in the D_1 protein of PSII	PSII electron transport and O_2 evolution in plants. Also inhibits electron transfer in photosynthetic bacteria
Atrazine (triazine)		Q_B-binding site in D_1 protein	PSII electron transport
Bromoxynil		Q_B site in D_1 protein	PSII electron transport
Paraquat (methyl viologen)		Bound Fd acceptors of PSI	NADP reduction and PSI electron transport. Formation of superoxide at PSI and membrane disruption
Phosphinothricin		Glutamine synthase	Glutamine biosynthesis
			CO_2 assimilation

Fig. 5.8 Herbicide structures and targets.

Herbicides

Herbicides (or weed killers) are chemicals which preferentially inhibit the growth of weeds in comparison to crop plants when applied to a crop community. Selectivity of action of a herbicide is a function of the structure and biochemical action of the compound and will depend on the differences in uptake rate, metabolism and target site of the herbicide between the weeds and the crops.

There are different types of herbicides based on their structure and site of action (Fig. 5.8). PSII herbicides such as diuron (DCMU), atrazine, dinoseb etc. are analogues of plastoquinone and act by displacing the secondary electron acceptor Q_B from its binding site on the D_1 protein. Phenolic inhibitors act by uncoupling electron transport from photophosphorylation. Paraquat acts as a PSI electron acceptor, diverting electrons from bound Fe–S centres and forming superoxide radicals. The widely used herbicide phosphinothricin (Glufosinate) is an analogue of glutamate and acts by blocking glutamate synthetase activity, which in turn restricts CO_2 assimilation in light.

Plant molecular biochemists are now able genetically to alter the target sites of herbicide action to make the recipients more resistant or more sensitive to the herbicide (§8.4). For example, in atrazine-resistant mutants of certain plants, the serine residue at position 264 in the D_1 protein of the wild type has been found to be substituted by a glycine. This is the result of a mutation of adenine in the PSbA gene (coding the D_1 protein) of the wild type to a guanine in the mutant. It is now possible to replace the Ser 264 by Gly by site-directed mutagenesis and to re-introduce the altered gene to engineer atrazine-resistance in plants.

Many herbicides that inhibit photosynthesis in plants also inhibit photosynthesis in bacteria. Herbicide-resistant mutants have been isolated for *Rhodobacter sphaeroides* and *Rhodopseudomonas viridis*. The comparative properties of the reaction centres from the mutant strains and wild types have enabled the biochemical identification of the binding sites of herbicides and some other electron transport inhibitors on the bacterial reaction centre (for example, the binding site of Q_B on D_1 protein).

6

Carbon dioxide fixation: the C₃ and C₄ pathways

In the previous chapter we have seen that $NADPH_2$ and ATP are produced in the light phase of photosynthesis. The fixation of CO_2 then takes place in the dark phase using the 'assimilatory power' of $NADPH_2$ and ATP. In this chapter we shall examine in some detail the reactions involved in the reduction of CO_2 to the level of carbohydrate since the reaction mechanisms and experimental techniques, so clearly worked out by Calvin and his co-workers from 1946 onwards, are some of the most important in modern biology. For his work in elucidating the path of carbon in photosynthesis, Calvin received the Nobel Prize for Chemistry in 1961.

6.1 Experimental techniques

When the long-lived isotope of carbon ^{14}C became available in 1945 its use, coupled with two-dimensional paper chromatography developed a few years earlier, enabled experiments to be devised to investigate the pathway of photosynthetic $^{14}CO_2$ fixation. The unicellular green algae *Chlorella* and *Scenedesmus* were used in the experiments because of their biochemical similarity to higher green plants and because they could be grown under uniform conditions and subsequently very quickly killed in the short-time experiments used.

Three main types of experiments were performed to obtain the evidence required to postulate the detailed reactions of the cycle.

(a) Exposure of the photosynthesizing algae to $^{14}CO_2$ for different lengths of time. At the shortest times only the initial products will be

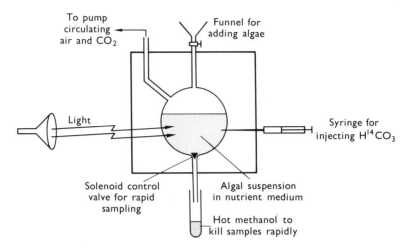

Fig. 6.1   Schematic representation of apparatus for studying $^{14}CO_2$ fixation in photosynthesizing algae.

radioactive. In this way phosphoglyceric acid (PGA) was identified as the primary carboxylation product; end-products such as sucrose became radioactive much more slowly.

(b) Determination of the position of radioactivity within the labelled compounds. In this way the details of the interconversions of sugar phosphate to regenerate the specific sugar phosphate which accepts the $^{14}CO_2$ molecule and the mechanism of synthesis of sugars and other compounds were worked out.

(c) Alteration of the external conditions, e.g. changing from light to dark, or changing from high to very low CO_2 concentrations, to see whether the cycle intermediates behave in a predictable manner.

The techniques employed are pictured in Figs 6.1, 6.2 and 6.3. Figure 6.1 is a diagram of the apparatus used for obtaining extracts of algae which have been photosynthesizing with $^{14}CO_2$. The algae are suspended in a nutrient medium through which air and CO_2 are bubbled, with the pH of the whole suspension maintained constant. The control valve allows rapid removal of samples into methanol which immediately stops all reactions from proceeding further. Figure 6.2 illustrates the further handling of the inactivated algae. The killed algal extract is concentrated by vacuum and then applied directly to a chromatogram paper which is developed with different solvents in two directions at right-angles to each other. The radioactivity of

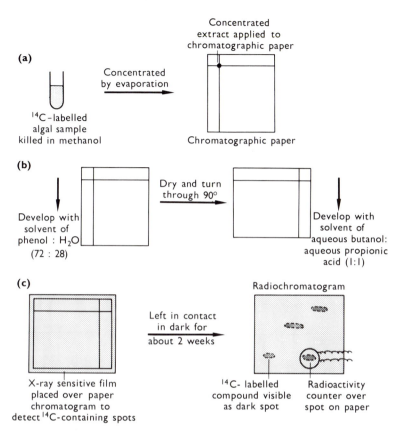

Fig. 6.2   Detection of the products of $^{14}CO_2$ fixation by algae after brief periods of illumination by the use of paper chromatography and autoradiography.

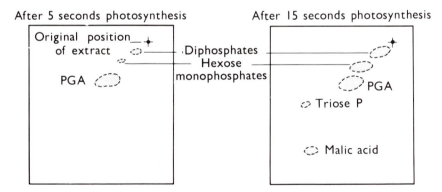

Fig. 6.3   Radioautograms of the photosynthetic products from $^{14}CO_2$ added for short periods of time to illuminated algae.

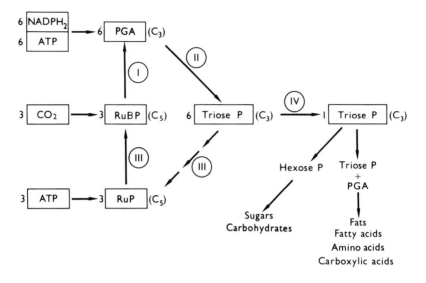

Fig. 6.4   Diagram of CO_2 fixation cycle. RuBP, ribulose bisphosphate; PGA, phosphoglyceric acid; Triose P, phosphoglyceraldehyde; RuP, ribulose-5-phosphate.

the compounds which are separated by this two-dimensional chromatography is measured – their location being known from radioautograms of the kind shown in Fig. 6.3. The two radioautograms in Fig. 6.3 show the compounds which contain ^{14}C in extracts of *Chlorella* which had been photosynthesizing for 5 and 15 seconds. It is seen that PGA, triose phosphate and sugar phosphates are formed very rapidly; sucrose, organic acids and amino acids are only formed after longer photosynthesizing times. The combination of radioactive CO_2 and two-dimensional chromatography is seen to be a very sensitive technique for detecting and quantitatively estimating the products of photosynthesis.

6.2   The photosynthetic carbon reduction (Calvin) cycle

The fixation of CO_2 to the level of sugar (or other compounds) can be considered to occur in four distinct phases, as is shown in Figs. 6.4 and 6.5.

I.   *Carboxylation phase*   This phase consists of a reaction whereby CO_2 is added to the 5-carbon sugar, ribulose bisphosphate, to form two

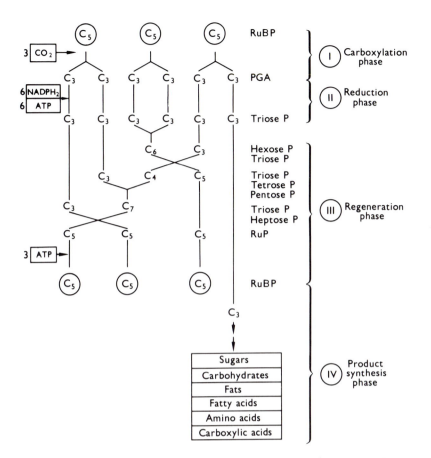

Fig. 6.5 Summary of reactions of photosynthetic CO_2 fixation. (For abbreviations, see text.)

molecules of PGA as follows:

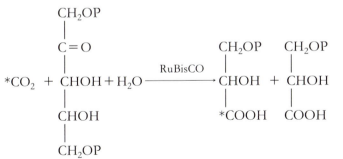

Ribulose bisphosphate (RuBP)   **2 × Phosphoglyceric acid (PGA)**

This reaction is catalyzed by the enzyme ribulose bisphosphate carboxylase (RuBisCO).

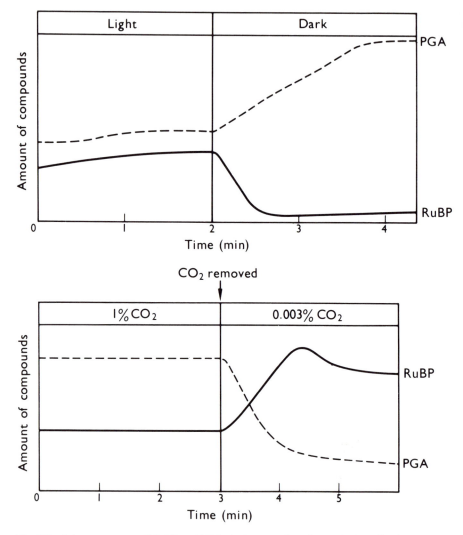

Fig. 6.6 Interconversions of RuBP and PGA, during experiments on photosynthesis.

The evidence in support of this scheme is shown very clearly in Fig. 6.6. On illumination, RuBP and PGA increase up to a certain level which is the so-called 'steady state' level in the photosynthesizing algae. When the light is switched off, the RuBP content drops immediately (as light is needed for its synthesis), while the level of PGA rises – two molecules of PGA are formed from every molecule of RuBP which disappears. Also in Fig. 6.6 is shown the effect of changing from high to very low CO_2 concentrations. A 'steady state' level is achieved at 1% CO_2 but when the CO_2 level is suddenly decreased to 0.003% (with the light still on), the level of PGA drops quickly as there is insufficient CO_2 to fix; however, the level of RuBP increases since very little of it can be used to fix CO_2 into PGA, while it can still be formed in the light.

II.   *Reduction phase*   PGA formed by the addition of CO_2 to RuBP is essentially an organic acid and is not at the energetic level of a sugar. In order for PGA to be converted to a 3-carbon sugar (triose P) the energy in the 'assimilatory power' of $NADPH_2$ and ATP must be used.

The reaction is in two steps: first, ATP-dependent phosphorylation of PGA at the COOH group to form 1,3-diphosphoglyceric acid and ADP, and second, reduction of the 1,3-diphosphoglyceric acid to phosphoglyceraldehyde by $NADPH_2$, releasing orthophosphate (Pi). The two steps can be summarized as follows:

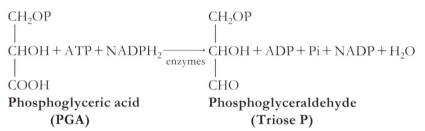

Phosphoglyceric acid   **Phosphoglyceraldehyde**
      (PGA)   **(Triose P)**

The two reactions are catalyzed by the enzymes phosphoglycerate kinase and glyceraldehyde 3-phosphate dehydrogenase respectively.

It is seen that the reducing power of $NADPH_2$ is used to change the acid group of PGA to an aldehyde group of the triose P; ATP is required to provide the extra energy in order to accomplish this step but the Pi of ATP is not incorporated into the triose P. Both of the enzymes involved in the two steps have been shown to be present in isolated chloroplasts.

Once the CO_2 has been reduced to the level of the 3-carbon sugar, triose P, the energy-conserving part of photosynthesis has been accomplished. What is required thereafter is to regenerate the initial CO_2 acceptor molecule, ie. ribulose bisphosphate, in order for the CO_2 fixation to continue again and again (regeneration phase) and to change the triose P to more complex sugars, carbohydrates, fats and amino acids (product synthesis phase).

III.   *Regeneration phase*   The RuBP is regenerated for further CO_2 fixation reactions by a complex series of reactions involving 3-, 4-, 5-, 6- and

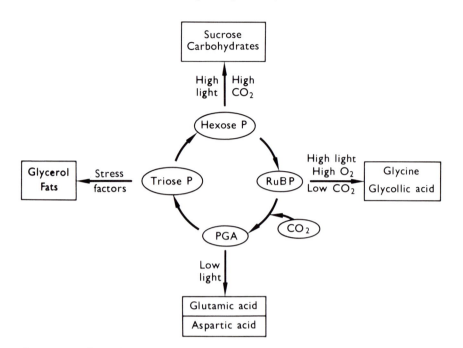

Fig. 6.7   Conditions favouring formation of secondary products in photosynthesis.

7-carbon sugar phosphates, which is depicted in summary form in Fig. 6.5. The details of the reactions are not important here; suffice it to say that all the reactions and the enzymes involved have been studied in some detail by various groups of research workers.

IV.   *Product synthesis phase*   End-products of photosynthesis are considered primarily to be sugars and carbohydrates but fats, fatty acids, amino acids and organic acids have also been shown to be synthesized in photosynthetic CO_2 fixation. Many details of these synthesis reactions are known but again they do not concern us directly. What is, however, interesting is that the different end-products seem to be formed under different conditions of light intensity, CO_2 and O_2 concentration, as is depicted in Fig. 6.7. Much research is now being aimed at working out the synthetic reactions involved in the formation of these end-products because an understanding of the reactions and the conditions favouring them may eventually enable us to induce plants to synthesize more or less sugars, fats or amino acids, by providing the required growth conditions.

6.3   Structure – function relationships

In order to study CO_2 fixation in isolated chloroplasts, they must be isolated rather carefully so as to preserve all the components of the reactions involved. Arnon's laboratory showed in 1954 that this could be accomplished with all the CO_2 fixation products being identified. However, the rates of fixation were less than one-tenth of those observed in leaves. Early in the 1970s, Walker's and Bassham's laboratories were able, by using very careful isolation media and procedures, to obtain chloroplasts capable of CO_2 fixation rates approaching those in whole leaves. Again, the products of photosynthesis were the same as those observed by Calvin's group in whole algae and by Arnon's laboratory in the early isolated chloroplasts.

6.4   Energetics of CO_2 fixation

If we look at the overall and constituent equations of photosynthesis again, we will be able to examine the energy-conserving and energy-expending parts of the carbon fixation cycle.

The general equation for formation of glucose can be represented by:

(i) $CO_2 + H_2O \rightarrow [CH_2O] + O_2$ $\Delta G = +48 \times 10^4$ J (114 kcal)

This means that 48×10^4 joules of energy are required to fix one mole of CO_2 to the level of glucose. This large positive value necessitates a large input of energy.

We have already seen that this energy is derived from the light phase of photosynthesis and can be represented by the 'assimilatory power' of $NADPH_2$ and ATP. In order to fix one CO_2 molecule, two molecules of $NADPH_2$ and three of ATP are required (see Figs. 6.4 and 6.5).

The energy present in the $NADPH_2$ and ATP can be represented as follows:

(ii) $2NADPH_2 + O_2 \rightarrow 2NADP + 2H_2O$ $\Delta G = -44 \times 10^4$ J (-105 kcal)

(iii) $3ATP + H_2O \rightarrow 3ADP + 3Pi$ $\dfrac{\Delta G = -9.2 \times 10^4 \text{ J}}{= -53.2 \times 10^4 \text{ J}}$ (-22 kcal)

This energy is sufficient to reduce one CO_2 molecule to the level of glucose with about 5×10^4 joules (13 kcal) to spare ($ii + iii - i$).

In summary, we can present the equations as:

$CO_2 + H_2O + 2NADPH_2 + 3ATP \longrightarrow [CH_2O] + O_2 + 2NADP + 3ADP + 3Pi$
$\Delta G = -5.2 \times 10^4$ J (-13 kcal)

Thus we see that photosynthesis is essentially a reductive process since 83% $(44 \times 10^4/53.2 \times 10^4) \times 100$ of the energy required to fix a molecule of CO_2 is derived from the strong reducing agent $NADPH_2$, which has a redox potential of -0.34 V. The redox potentials of sugars can be thought to be approximately -0.43 V; so the ATP is required to fix the CO_2 to this lower redox value.

We know that the redox potential of H_2O/O_2 is $+0.82$ V; so the overall change in redox potential is 1.25 V ($+0.82$ to -0.43 V). This can be converted into terms of energy using the equation:

$$\Delta G = -nF\Delta E$$
$$= -(4)(9.64 \times 10^4)(1.25) = -48.2 \times 10^4 \text{ J (116 kcal)}$$

where n = number of electrons ($= 4$ electrons per molecule of oxygen)
         F = the Faraday ($= 9.64 \times 10^4$ J per volt equivalent)
         ΔE = difference in redox potential

It is seen that this ΔG value is very close to that for fixing one molecule of CO_2 (48×10^4 J = 114 kcal; equation (*i*) above) and shows quite nicely the interconversion of energy in terms of joules and redox potentials.

Lastly, we can discuss the quantum efficiency of CO_2 fixation. Each mole quantum of red light at 680 nm contains 17.61×10^4 J of energy (Table 1.1). Thus, at least three ($48 \times 10^4/17.61 \times 10^4 = 2.7$) mole quanta of 680 nm light will be required for one CO_2 molecule to be fixed. However, experimentally it is found that 8–10 quanta of absorbed light are required for each molecule of CO_2 fixed or O_2 evolved. From our knowledge of non-cyclic photosynthetic phosphorylation we deduce that there are two different light reactions required to reduce NADP with the electrons from H_2O:

$$2NADP + 2H_2O \xrightarrow[\substack{\text{2 light reactions} \\ \text{chloroplasts}}]{4e^-} 2NADPH_2 + O_2$$

Thus we need at least 8 quanta (4 quanta per 4e (1 O_2 molecule) $\times$ 2 light reactions) to reduce NADP and produce the necessary ATP at the same time.

Nevertheless, photosynthetic CO_2 fixation itself is only about 30% efficient (2.7 quanta/8–10 quanta) as we can measure it. Taken in conjunction with an average efficiency of less than 1% for whole plants capturing and utilizing photosynthetically active sunlight (see Chapter 1), this reinforces the concept that these energy exchanges are necessary but wasteful and could be improved in artificial photosynthetic systems.

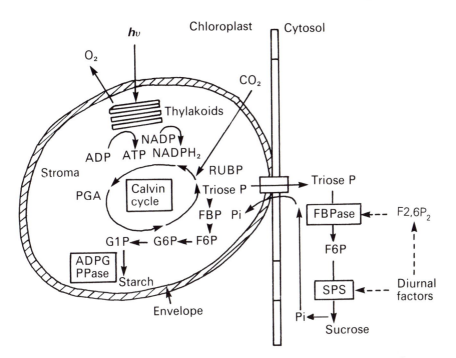

Fig. 6.8   Regulation of sucrose and starch synthesis in leaves. ADPG PPase, ADP glucose pyrophosphorylase; F6P, fructose 6-phosphate; F2,6P$_2$, fructose 2,6-bisphosphate; G1P, glucose 1-phosphate; Pi, orthophosphate; SPS sucrose phosphate synthetase. The phosphate translocator is situated in the chloroplast envelope. (After Stitt & Quick, 1989.)

6.5   Sucrose and starch synthesis

The net outcome of the Calvin cycle is the generation of one molecule of triose phosphate (DHAP or PGA) from three molecules of CO_2 during every three turns of the cycle. The overall process can be summarized as:

$$3CO_2 + RuBP \xrightarrow[\text{enzymes (stroma)}]{6NADPH_2 + 9ATP \text{ (thylakoids)}} 1 \text{ triose P} + RuBP$$

The triose P is used for starch synthesis within the chloroplast or for sucrose synthesis in the cytosol. The triose phosphates are exported from the chloroplasts to the cytosol in exchange for orthophosphate (Pi). This exchange is mediated by the phosphate translocator protein situated in the chloroplast envelope membrane (Fig. 6.8). In the cytoplasm the triose phosphates combine to form fructose 1,6-bisphosphate (FBP), which is subsequently hydrolyzed to fructose 6-phosphate by the enzyme fructose bisphosphatase. Fructose 6-phosphate is isomerized to glucose 1-phosphate via glucose 6-phosphate.

The synthesis of sucrose involves the participation of uridyl (U) phosphates and the following steps catalyzed by cytoplasmic enzymes.

Glucose 6-phosphate $\longrightarrow$ glucose 1-phosphate (Phosphoglucoisomerase)
Glucose 1-phosphate + UTP $\longrightarrow$ UDP-glucose + pyrophosphate

There are two pathways for conversion of UDP-glucose to sucrose. In plants such as sugar cane:

UDP-glucose + fructose 6-phosphate $\longrightarrow$ sucrose 6-phosphate + UDP
Sucrose 6-phosphate + H_2O $\longrightarrow$ sucrose + Pi

In some other plants the following pathway exists:

UDP-glucose + fructose $\longrightarrow$ UDP + sucrose

Under conditions where the CO_2 fixation rate in chloroplasts exceeds the rate at which triose phosphates can be converted to sucrose in the cytoplasm, synthesis of starch (a glucose polymer) occurs in the chloroplast stroma. The triose phosphates are converted to glucose 1-phosphate which then reacts with ATP as:

Glucose 1-phosphate + ATP $\longrightarrow$ ADP glucose + pyrophosphate

This reaction is catalyzed by the enzyme ADP-glucose pyrophosphorylase.

$$\text{ADP-glucose} + [\text{glucose}]_n \xrightarrow{\text{starch synthase}} \underset{\text{starch}}{[\text{glucose}]_{n+1}} + \text{ADP}$$

The starch synthesized in the chloroplasts is broken down, to sugars or sugar phosphates (depending on the availability of Pi), and utilized by the plant during periods of darkness or limited photosynthesis.

Regulation of sucrose and starch synthesis   For every six molecules of triose phosphates formed during the Calvin cycle, one molecule is made available for export to the cytosol for sucrose synthesis; the remaining five being needed to regenerate RuBP (see Figs. 6.4 and 6.5).

In order for CO_2 assimilation to proceed at good rates and continuously in light, it is essential that the rate of sucrose synthesis in the cytoplasm is maintained in such a way that: (1) the rate of export of triose-P from the stroma does not limit its availability for the continuation of the cycle, and (2) sufficient Pi is released from sucrose synthesis (final step) for import back into the chloroplast stroma to allow phosphorylation of the Calvin cycle intermediates.

The partitioning of triose phosphate between sucrose synthesis in the cytoplasm and starch formation in the chloroplasts is regulated by a number of factors. The major reactions that control sucrose synthesis have been formulated by Heldt and Stitt in Germany from studies of equilibrium constants of the enzymatic steps, the use of inhibitors, and the use of mutants lacking a particular enzyme in the synthesis pathway. It appears that the two key enzymes that regulate sucrose synthesis are the cytosolic enzymes fructose bisphosphatase (FBPase) and sucrose phosphate synthase (SPS). The FBPase which catalyzes the hydrolysis of fructose 1,6-bisphosphate is inhibited by fructose 2,6-bisphosphate, a cytosolic metabolite, whose concentration in leaves varies with the incident light intensity. The activity of sucrose phosphate synthase is regulated by light (diurnal) and by metabolites (G-6P and Pi) of the sucrose synthesis pathway, possibly through reversible modification of the protein.

An enzyme controlling starch biosynthesis in the chloroplast stroma is ADP glucose pyrophosphorylase. The catalytic rate of this enzyme is stimulated by PGA and inhibited by Pi; thus the rate of starch synthesis is dependent on the [PGA]:[Pi] ratio in the chloroplast stroma.

Many C$_4$ plants, when grown under natural conditions, synthesize sucrose in the mesophyll cells and starch in the bundle-sheath chloroplasts. Sucrose, being a neutral (non-electrolyte), non-reducing sugar which is highly soluble in water, serves as a useful metabolite regulating the osmotic and water balance between the various cellular constituents of the plant. It is also the main storage sugar of plants (Fig. 6.9).

6.6   The C$_4$ (Kortschak, Hatch–Slack) pathway of CO$_2$ fixation

Many tropical grasses and plants such as sugar cane and maize are able to fix CO$_2$ initially into 4-carbon compounds like oxaloacetate, malate and aspartate, in addition to the CO$_2$ fixation which occurs via the Calvin C$_3$ cycle. The formation of C$_4$ acids as initial CO$_2$-reduction products was first observed in sugar cane leaves (§2.6).

As already mentioned (§3.4), the leaves of these C$_4$ plants possess two types of chloroplasts, mesophyll and bundle-sheath. The stomata of the C$_4$

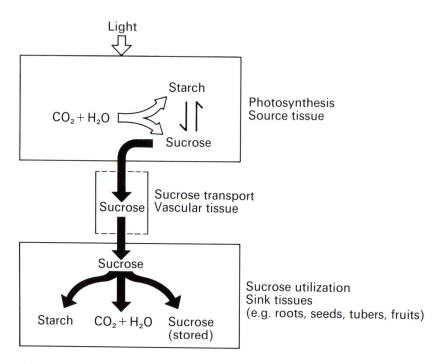

Fig. 6.9　Simplified scheme showing the synthesis and distribution of sucrose in plants. (Courtesy: B. B. Buchanan, University of California, Berkeley.)

plants are usually located in such a way that the substomatal cavity is immediately adjacent to the mesophyll cell chloroplasts. The CO_2 which diffuses into the leaf through the stomata enters the mesophyll cytoplasm where it reacts with phosphoenol pyruvate (PEP) to form oxaloacetate in the presence of the enzyme PEP carboxylase.

$$CO_2 + PEP[CH_2 = \overset{\overset{\displaystyle OP}{\displaystyle |}}{C} - COOH] \xrightarrow[\text{carboxylase}]{\text{PEP}} \text{oxaloacetate}$$

There is a high concentration of PEP carboxylase in the mesophyll cells of C_4 species and thus CO_2 can be readily fixed down to low concentrations. The oxaloacetate is subsequently reduced by $NADPH_2$, formed by the normal light reactions, to malate. This reaction is catalyzed by an NADP-specific malate dehydrogenase, another enzyme found in abundance in the mesophyll cells of C_4 plants and whose activity is light regulated. Radio-activity-labelling experiments using $^{14}CO_2$ have shown that more than 90%

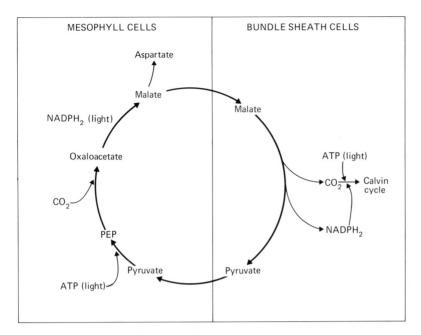

Fig. 6.10   CO$_2$ fixation scheme in C$_4$ plants.

of the radioactivity is fixed in C$_4$ acids in 1 second. The malate is transported to the bundle-sheath cells where it is decarboxylated to pyruvate and CO$_2$, and the CO$_2$ thus released is then used for sugar and starch production via the Calvin C$_3$ cycle. The malate can also function as a constituent of the Kreb's cycle or can be aminated to aspartate and form a constituent of the amino acid pool. A simplified scheme of CO$_2$ fixation by C$_4$ plants is given in Fig. 6.10.

The rate of photosynthetic CO$_2$ fixation by C$_4$ plants is not affected by atmospheric concentrations of O$_2$ (which is high) and CO$_2$ (low), both factors which normally enhance the photorespiration rate of C$_3$ plants. The water-use efficiency, i.e. the ratio of the mass of CO$_2$ assimilated to water transpired, in C$_4$ plants is often twice that of C$_3$ species; the specialized Krantz anatomy is a contributing factor for this trait (§3.4). Also, salinity tolerance is a common feature of many C$_4$ species. All these traits allow C$_4$ plants to survive in dry and saline habitats.

However, compared to C$_3$ plants, the C$_4$ species require an extra two or three molecules of ATP for fixation of one CO$_2$ molecule, the extra ATPs being consumed in the phosphorylation of pyruvate to PEP. The C$_4$ pathway is thus an additional step in photosynthesis which can be

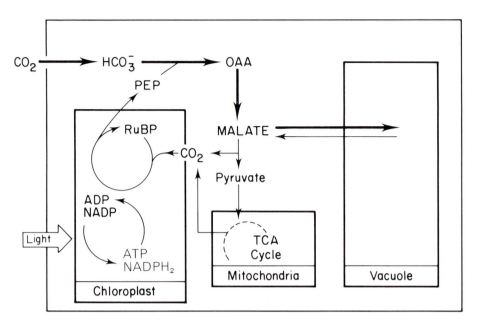

Fig. 6.11 Flow of carbon in CAM plants. Thick arrows indicate reactions which predominate in the dark; thin arrows indicate reactions which occur in the light. OAA, oxaloacetate; PEP, phosphoenol pyruvate; RuBP, ribulose bisphosphate; TCA, tricarboxylic acid.

considered as an *ATP-dependent* CO_2 pump. This energetically wasteful appendage in C_4 plants is compensated by their high rates of photosynthesis and associated sucrose formation. Carbon assimilation in C_4 photosynthesis is not fully saturated, even at high PFDs. Modelling experiments have predicted a steady state CO_2 concentration of at least 70 μM in the bundle-sheath cytosol. This is about 20 times the steady state CO_2 level in the adjacent mesophylls during photosynthesis and is high enough to prevent any oxygenase activity by the RuBisCO causing photorespiratory CO_2 loss.

6.7   Crassulacean acid metabolism: CAM species

Many succulent plants growing in arid environments *fix CO_2 in the dark* to the C_4 acids oxaloacetic and then malic – the phenomenon was investigated extensively in the Crassulaceae and termed *Crassulacean acid metabolism* (CAM). CAM is widespread in the angiosperm families Agavaceae,

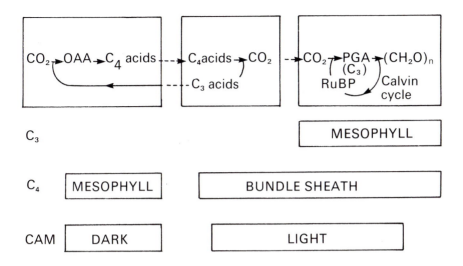

Fig. 6.12   The principal features of photosynthetic CO_2 assimilation in C_3, C_4 and CAM species, illustrating the spacial segregation of the two carboxylations in the C_4 and the temporal segregation in the CAM plants. (After Lawlor, 1987. See Lawlor, 1993, for details.) See Fig. 6.11 for abbreviations.

Bromeliaceae, Cactaceae, Crassulaceae, Euphorbiaceae, Liliaceae, Orchidaceae, etc. CAM plants normally close their stomata during the day to prevent water loss. Their stomata open at night. CO_2 enters the leaves and combines with PEP (a product of starch metabolism) to form oxaloacetic acid in the presence of the enzyme PEP carboxylase which is found in the cytoplasm of the leaf cells. The oxaloacetate is reduced by malic dehydrogenase to malic acid, which accumulates in the leaf vacuoles (Fig. 6.11). During the day the stomata become closed, the malate is transported to the cytoplasm where it is decarboxylated by a malic enzyme to yield pyruvate and CO_2. The CO_2 thus released enters the chloroplasts where it is fixed to sugars by the photosynthetic Calvin C_3 cycle. Thus in CAM plants the fixation of CO_2 to malate at night and its decarboxylation to CO_2 and pyruvate during the day are separated in time, whereas in C_4 plants the two phases are separated spatially where the primary carboxylation to C_4 acids occurs in the mesophyll and the decarboxylation and secondary CO_2 fixation to PGA reactions in the bundle-sheath. Given adequate water, CAM plants may behave like C_3 species.

The three types of CO_2 reduction pathways are shown in Fig. 6.12, and some of the distinguishing characteristics of C_3, C_4 and CAM in Table 6.1.

Table 6.1. *General characteristics of C_3, C_4 and CAM plants*

	C_3	C_4	CAM
1.	Typically temperate species, e.g. spinach, wheat, potato, tobacco, sugar beet, soya bean, sunflower, eucalyptus, pines.	Typically tropical or semi-tropical species, e.g. maize, sugar cane, *Amaranthus, Sorghum,* savannah grasses. Plants adapted to high light, high temperatures and also semi-arid environments.	Typically arid zone species, e.g. cacti, orchids, *Agave,* succulent plants.
2.	Moderately productive. Yields of 30 t (tonnes) dry weight per hectare (2.47 acres) posssible e.g. eucalyptus.	Highly productive, 120 t wet weight per hectare for sugar cane is possible.	Usually very poorly productive. (Pineapple is highly productive.)
3.	Cells containing chloroplasts do not show Kranz-type anatomy and generally lack peripheral reticulum. Only one type of chloroplast.	Kranz-type anatomy and peripheral reticulum are essential features. Often have two distinct types of chloroplasts.	Lack Kranz anatomy and peripheral reticulum. Only one type of chloroplast
4.	Initial CO_2 acceptor is ribulose bisphosphate (RuBP), a 5-carbon sugar.	Initial CO_2 acceptor is phosphoenol pyruvate (PEP), a 3-carbon acid.	CO_2 acceptor is PEP in the dark and RuBP in the light.
5.	Initial CO_2 fixation product is the 3-carbon acid phosphoglycerate.	Initial CO_2 fixation product is the 4-carbon acid oxaloacetate.	CO_2 fixation products are oxaloacetate in the dark and phosphoglycerate in light.

6.	Only one CO_2 fixation pathway.	Two CO_2 fixation pathways separated in space.	Two CO_2 fixation pathways separated in time.
7.	High rates of glycollate synthesis and photorespiration	Low rates of glycollate synthesis; no photorespiration.	Same as C_4.
8.	Low water use efficiency and salinity (ion) tolerance.	High water use efficiency and salinity tolerance.	Same as C_4.
9.	Photosynthesis saturates at 1/5 full sunlight.	Do not readily photosaturate at high light.	Same as C_4.
10.	High (30 to 80 p.p.m.) CO_2 compensation point.	Low (<10 p.p.m.) CO_2 compensation point.	Low (<5 p.p.m.) CO_2 compensation point.
11.	Open stomata by day.	Open stomata by day.	Open stomata by night.

6.8   Light-coupled reactions of chloroplasts other than CO_2 fixation

In addition to CO_2 assimilation, there are many other 'dark' reactions localized in the chloroplast stroma (and also envelope) which utilize ATP, $NADPH_2$, reduced Fd, and sugar phosphates generated by photosynthetic electron transport. Some of these light-coupled reactions which are essential for the synthesis of organic material by plants are outlined here.

(a) Assimilation of nitrogen   Nitrate is the most important nitrogen source for plants. The first step in the assimilation of nitrate is its reduction to nitrite, catalyzed by the enzyme nitrate reductase, the electrons being donated by reduced NAD or NADP.

$$NO_3^- + NAD(P)H_2 \longrightarrow NO_2^- + NAD(P) + H_2O$$

Nitrate reductase is present in roots as well as in leaves. Isolated chloroplasts do not reduce nitrate since the enzyme is present only in the cytosol. However, nitrate reduction in leaf extracts is stimulated in light due to an increased supply of reductants (possibly phosphoglyceraldehyde or malate) from the chloroplasts which generate $NADH_2$ in the cytosol ($NADPH_2$ produced in the chloroplasts cannot pass through the chloroplast envelope). Some species of blue-green algae contain a nitrate reductase which can take up electrons from reduced ferredoxin for the reduction of nitrate.

The enzyme nitrite reductase, which catalyzes the reduction of nitrite to ammonia, is located in the stroma. Reduced ferredoxin is the electron donor for this reaction.

$$NO_2^- + 6e^- + 8H^+ \longrightarrow NH_4^+ + 2H_2O$$

Since ammonia is toxic to the plant, it is immediately assimilated as glutamine via an ATP-dependent reaction catalyzed by glutamine synthetase:

$$NH_3 + glutamate + ATP \longrightarrow glutamine + ADP + Pi$$

Many species of blue-green algae can fix atmospheric nitrogen to ammonia. This nitrogen assimilation is catalyzed by the enzyme nitrogenase and utilizes electrons from reduced ferredoxin and energy from ATP – both products of photosynthetic electron transport. Many of the biochemical steps in photosynthetic nitrogen assimilation were discovered by Losada's research group in Seville.

(b) Assimilation of sulphate   Reductive assimilation of sulphate occurs mainly in the leaves – sulphate probably enters the chloroplasts via a

sulphate translocator in the chloroplast envelope. In isolated chloroplasts, sulphate assimilation is light dependent and requires both reduced ferredoxin and ATP. Sulphate is initially reduced to sulphite (with the consumption of ATP) and further to sulphide via sulphite reductase, a ferredoxin-dependent enzyme localized in the chloroplast. Cysteine synthetase, another chloroplast enzyme, immediately assimilates sulphide as the amino acid cysteine, which is then further metabolized to other biological sulphur compounds such as glutathione, methionine, Coenzyme A (CoA), sulpholipids, etc.

(c) Fatty acid biosynthesis   Chloroplasts are the main site of fatty acid synthesis in leaves. Fatty acid biosynthesis is catalyzed by the fatty acid synthases in the stroma and requires photosynthetically produced ATP and NADPH$_2$. The primary precursor is acetyl CoA (possibly derived from acetate) and the main products are the saturated fatty acids, palmitic (C_{16} and stearic (C_{18}), and the unsaturated fatty acid, oleic ($C_{18:1}$).

(d) Oxygen exchange   Oxyen is a by-product of photosynthetic water splitting leading to CO$_2$ fixation; the concentration of O$_2$ in the chloroplasts in the light is always higher than that of the air surrounding the leaf. Accumulation of O$_2$ will be toxic to the chloroplasts since it can cause oxidation of membrane lipids and inhibition of enzymes of the CO$_2$ fixation pathway such as ribulose bisphosphate carboxylase (by binding to the enzyme) and glyceraldehyde 3-phosphate dehydrogenase (by oxidation of − SH groups). Oxygen and its reduction products are metabolized in the chloroplasts by a number of pathways (Fig. 6.13). Photorespiration in C_3 plants (§6.9) is one such means of disposing of molecular O$_2$. Chloroplasts contain a high (millimolar) concentration of reduced glutathione (GSH, glutamyl cysteinyl glycine), which can react with O$_2$:

$$2GSH + \tfrac{1}{2}O_2 \longrightarrow 2GSSG + H_2O$$

The oxidized glutathione is reduced back to GSH by NADPH$_2$, a reaction catalyzed by glutathione reductase, a stromal enzyme:

$$GSSG + NADPH_2 \longrightarrow 2GSH + NADP$$

Excited chlorophyll molecules in their triplet state (Fig. 4.1) can transfer their energy to the oxygen molecule generating an excited oxygen species known as singlet oxygen. Singlet oxygen can damage thylakoid membrane structure by oxidizing polyunsaturated fatty acids to lipid peroxides. This oxidative damage is prevented by the removal of singlet oxygen by

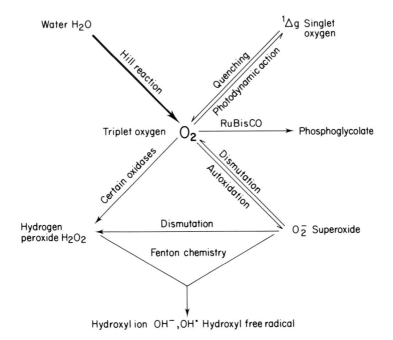

Fig. 6.13   Major pathways of oxygen metabolism in the active chloroplast.

reductants in chloroplasts such as α-tocopherols, carotenoids, ascorbate, and GSH.

Monovalent reduction of O_2 by photoreduced PSI electron acceptors (reduced Fe–S centres and ferredoxin) has been demonstrated in intact chloroplasts and in leaves at rates varying between 3% and 27% of total electron transport, dependent on the irradiance and CO_2 concentration.

The primary product of such O_2 reduction by reduced ferredoxin is superoxide (O_2^-), a highly active free radical species:

$$O_2 + Fd_{red} \longrightarrow Fd_{ox} + O_2^-$$

The superoxide thus generated is converted to H_2O_2 via two different pathways:

(i) $O_2^- + 2H^+ + Fd_{red} \longrightarrow Fd_{ox} + H_2O_2$
(ii) $2O_2^- + 2H^+ \longrightarrow O_2 + H_2O_2$

The second reaction, dismutation of superoxide, is catalyzed by the Cu–Zn-containing *superoxide dismutase* enzyme occurring in the chloroplast stroma.

Hydrogen peroxide is a toxic molecule and it is reduced to water by the catalytic action of metal ions and also by ascorbate present in high concentrations (up to 50 mM) in the chloroplasts:

$$2H^+ + H_2O_2 + ascorbate \xrightarrow[\text{peroxidase}]{\text{ascorbate}} 2H_2O + dehydroascorbate$$

As already mentioned (§5.4), the photoreduction of O_2 by reduced ferredoxin is coupled to phosphorylation of ADP – this type of non-cyclic photophosphorylation is one of the major means of balancing the ATP: $NADPH_2$ ratio in the chloroplasts.

6.9   Photorespiration and glycollate metabolism

An active field of current research is the study of photorespiration (the light-stimulated release of CO_2 at rapid rates by leaves), which is quite different from the 'dark' evolution of CO_2 by mitochondrial respiration in leaves. Photosynthetic carbon oxidation in light was first reported by Decker in 1957. Plant species differ markedly in their rates of photorespiration: in some inefficient photosynthetic species it may be as high as 50% of net photosynthesis. Labelling experiments using the heavy oxygen isotope [18]O have shown that in plants with high photorespiration rates, one of the initial products containing the [18]O label is the 2-carbon (C_2) acid, glycollic acid, followed in time by glycine, serine and 3-phosphoglyceric acid [PGA], a Calvin cycle intermediate. The proposed metabolic pathway of glycollic acid oxidation which leads to CO_2 formation (during the conversion of glycine to serine) is shown in Fig. 6.14. There may be additional oxidation reactions producing CO_2. How is glycollic acid formed in the chloroplasts? The enzyme ribulose bisphosphate carboxylase (RuBisCO), which is the major protein of the chloroplast stroma, can bind CO_2 as well as O_2, i.e. the enzyme can function as a carboxylase and as an oxygenase (§8.10). When it functions as an oxygenase it catalyzes the oxygenation of RuBP to 2-phosphoglycollic acid and PGA. The phosphoglycollic acid is then hydrolyzed to phosphate and glycollic acid (Fig. 6.15).

Both CO_2 and O_2 compete for RuBP at the same site of the enzyme, so that high concentrations of CO_2 and low concentrations of O_2 favour carboxylation, whereas high concentrations of O_2 and low concentrations of CO_2 (as is found in the atmosphere) favour oxygenation and thus formation of phosphoglycollic acid. It has also been reported that higher

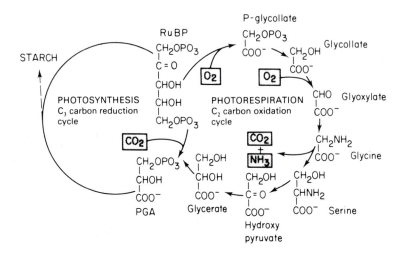

Fig. 6.14   Integrated carbon reduction and photorespiratory carbon oxidation cycle.

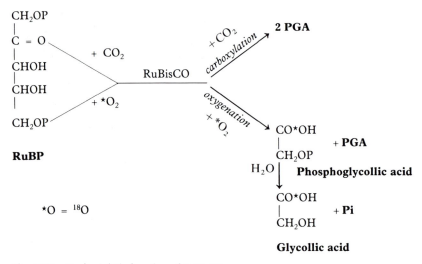

Fig. 6.15   Dual catalytic function of RuBisCO.

temperatures favour the oxygenation reaction by the enzyme. Glycollic acid can be formed by other metabolic pathways also in the chloroplast.

   The glycollate is exported from the chloroplast to the peroxisome (an adjoining organelle in the plant cell) where it is oxidized to glyoxylate and then aminated to glycine. The glycine gets transported to the mitochondrion, which is the third organelle participating in photorespiration. The source of photorespiratory CO_2 is the conversion of glycine to serine, which takes place in the mitochondrion.

$$2CH_2NH_2COOH + NAD + H_2O \longrightarrow CH_2OH - CHNH_2COOH + NADH_2 + CO_2 + NH_3$$

$\quad$ (glycine) $\qquad\qquad\qquad\qquad\qquad$ (serine)

The ammonia released is re-assimilated through the glutamate synthase–glutamyl oxalylgamma amino transferase enzyme (GS/GOGAT) coupled to glutamate in the chloroplasts. The glutamate is subsequently exported to the peroxisomes for transamination of glyoxylate to glycine.

The external symptoms of photorespiration are (1) the inhibition of photosynthesis by increased O_2, (2) the existence of a high CO_2 compensation point (30–50 ppm CO_2 at 25°C in air), and (3) the variation of the CO_2 compensation point in response to variations in O_2, light and temperature.

Plants belonging to the C_4 and CAM species do not show these symptoms, though the kinetic properties of RuBisCO isolated from these species are similar to those of this enzyme from C_3 plants or even from photosynthetic bacteria. In the C_4 plants most of the RuBisCO is compartmentalized in the chloroplasts of bundle-sheath cells, which are not themselves in direct equilibrium with the atmospheric CO_2 or O_2. The CO_2 concentration in the bundle-sheath cells may be much higher than that in the atmosphere since the CO_2 is produced *in situ* by the decarboxylation of malate which is imported from the mesophyll cells (see §6.6) – this allows carboxylation to compete more effectively with oxygenation for the enzyme-bound RuBP. Also, any CO_2 generated by photorespiration in the C_4 species could be trapped in the chloroplasts as a result of internal recycling by the PEP carboxylase of the mesophyll cells so that the loss of CO_2 to the atmosphere is prevented. The CAM plants synthesize a high level of malate at night. Since decarboxylation of malate and CO_2 fixation by the C_3 pathway in CAM species occurs in the day when the stomata are closed (see §6.7), the internal concentration of CO_2 in these plants may well be much higher than that of the external atmosphere, thus stimulating carboxylation over oxygenation.

6.10   Environmental factors affecting CO_2 assimilation by plants

In Chapter 2 we discussed how the external factors, viz. light, temperature, and ambient CO_2 concentration, affected photosynthesis, with particular reference to the green alga, *Chlorella*. Photosynthetic light-response curves

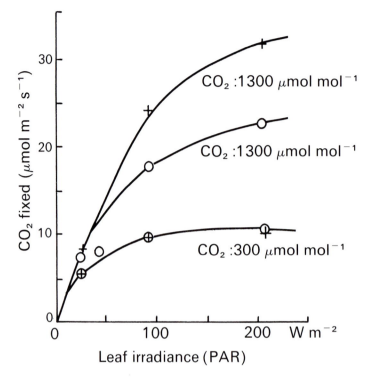

Fig. 6.16   Gross CO_2 assimilation rate of a cucumber leaf (a C_3 plant). 100 W m^{-2} PAR = approximately 400 μmol m^{-2} s^{-1} PPFD. (Courtesy: J. Goudriaan, Wageningen.) (○) 20°C; (+) 30°C.

for higher plants are more complex. There are differences between C_3 and C_4 species, and even in the same species the response varies depending on the PPFD in which the plants are cultivated. In leaves of 'shade' species or in shade-acclimatized leaves, net photosynthesis may be saturated at a PPFD of 100 μmol m^{-2} s^{-1}, whereas 'sun' leaves continue to increase their photosynthetic rates, albeit slowly, up to typical values of full sunlight provided there is sufficient CO_2, water and other nutrients available. At near-optimal temperatures and in saturating CO_2, all plants photosynthesize at the same rate in *low* light (Figs. 2.2 and 6.16) with a quantum requirement of approximately 9. Continued exposure of many species to very high irradiance can damage the photosynthetic system, due to photoinhibition (§8.13), and lower their productivity.

If the amount of light incident on a leaf surface is decreased gradually from that of full sunlight, eventually a PPFD value is reached for which there is no gain or loss of photosynthesis. This PPFD at which there is no net change in photosynthesis, usually measured as CO_2 exchange, is known as the *light compensation point* for the species. At 20°C and 350 ppm (350 μmol mol^{-1}) CO_2, the light compensation occurs usually at a PPFD of 8 to 15 μmol m^{-2} s^{-1}.

In a plant canopy most leaves are subjected to rapidly alternating periods of sun and shade because of sunflecks. Due to these sunflecks, a large fraction of CO_2 assimilation in canopies may be occurring under transient light conditions wherein the environmental and physiological constraints on photosynthetic CO_2 fixation may differ from those under steady state conditions. The sizes of sunflecks and their maximum light intensity vary with the canopy structure, which is dependent on factors such as the degree of dispersion of the foliage, the cumulative leaf area index etc. Direct measurements of light incidence and gas (O_2 and CO_2) exchanges in the field have shown that on clear days a high proportion (up to 80%) of the light received and CO_2 exchanged in the understories of forest canopies are attributable to sunflecks (see also §8.16).

The ambient CO_2 concentration at which the CO_2 released from leaves by respiration and photorespiration is exactly balanced by photosynthetic CO_2 assimilation is known as the *CO_2 compensation point*. The CO_2 compensation point is an indicator of the capacity of the plant to absorb the CO_2 from the environment and to assimilate it efficiently in light; the lower the compensation point, the better the CO_2 photoassimilation capacity. The CO_2 compensation point generally increases with increasing temperature (due to higher photorespiration) and decreasing light. It is very low (< 10 μmol mol^{-1}) for the C_4 plants, whereas for the C_3 species the values exceed 50 μmol mol^{-1}, measured at 25°C and in saturating light.

Plants can photosynthesize in habitats having a broad range of temperatures: from near to 0°C in the alpine areas to 50°C in the deserts. Temperature affects all the biochemical reactions of photosynthesis, and the responses to temperature are complex and depend upon the species, the ambient CO_2 levels, incident light energy etc. In the C_4 species, with practically no photorespiration, the quantum yield stays constant with temperature, whereas in the C_3 plants the quantum yield decreases with increase of temperature, reflecting a stimulation of photorespiration and a concomitant higher energy consumption per net CO_2 fixed.

The effects of temperature, light intensity and ambient CO_2 concentration on photosynthesis in a C_3 plant are illustrated in Fig. 6.16.

7

Bacterial photosynthesis

7.1 Classification

Photosynthetic bacteria are typically aquatic micro-organisms inhabiting marine and freshwater environments like moist and muddy soil, stagnant ponds and lakes, sulphur springs, etc. They are classified into four major types based on their pigment composition, membrane structure and metabolic requirements.

1. *Green bacteria.* These are subdivided into two families, viz. the Chlorobiaceae and the Chloroflexaceae. The Chlorobiaceae are strict anaerobes which grow by utilizing sulphide or thiosulphate as electron source, e.g. *Chlorobium limicola* and *Ch. thiosulfatophilum.* The Chloroflexaceae are facultative aerobes which can utilize reduced carbon compounds as electron donors for their growth, e.g. *Chloroflexus aurantiacus.* The two families are similar in their photosynthetic pigment composition and membrane fine structure but have different types of electron transfer pathway.

2. *Purple sulphur bacteria* (Chromatiaceae) which can use hydrogen sulphide as a photosynthetic electron donor, e.g. *Chromatium.*

3. *Purple non-sulphur bacteria* (Rhodospirillaceae) which are unable to use hydrogen sulphide and depend on the availability of simple organic compounds like alcohols and acids as electron donors, e.g. *Rhodomicrobium, Rhodobacter, Rhodopseudomonas* and *Rhodospirillum.*

4. *Heliobacteria. Heliobacterium chlorum* is a recently (1983) isolated, strictly anaerobic photosynthetic bacterium that contains a previously unknown type of bacteriochlorophyll, BChl *g,* as antenna and reaction

centre pigment. In addition to the pigment composition, the photosynthetic membrane architecture of *H. chlorum* is different from those of other photosynthetic prokaryotes, necessitating the introduction of a new type, viz. *Heliobacteria*, to describe members of this group of organisms.

When grown photosynthetically, all four types are strict anaerobes, i.e. grow only in the complete absence of oxygen. They cannot use water as a substrate and they do not evolve oxygen during photosynthesis.

7.2   Photosynthetic pigments and apparatus

The pigment systems of photosynthetic bacteria are slightly different from those of plants and algae. The chlorophyllous pigments of bacteria are called *bacteriochlorophylls*; six classes of bacteriochlorophylls (BChl *a*, BChl *b*, BChl *c*, BChl *d*, BChl *e* and BChl *g*) have been characterized. These bacteriochlorophylls are very similar to chlorophylls *a* and *b* but differ in the nature of the side chains attached to the carbon atoms 2, 3, 4, 5, 7 and 10 shown for chlorophyll in Fig. 3.7. In addition, a magnesium-less bacteriochlorophyll which is called bacteriopheophytin is found in the reaction centre of all photosynthetic bacteria. The principal carotenoids of photosynthetic bacteria are also slightly different chemically from the algal carotenoids. The nature of some of the pigments found in the photosynthetic bacteria and their growth requirements are given in Table 7.1. The absorption spectra of two typical bacteria are shown in Fig. 7.1.

The photosynthetic apparatus in purple and green bacteria are morphologically different and both are distinct from the photosynthetic unit found in chloroplasts. The action spectra of BChl fluorescence in purple bacteria indicate that light energy absorbed by carotenoids and shortwave BChl bands is transferred to the longest wavelength BChl (which absorb at 870, 890 and 960 nm) before being used for photosynthesis. From measurements of substrate (carbon) assimilation and photosynthetic phosphorylation by suspensions of purple bacteria, during flashing light experiments, Clayton (Cornell, New York) estimated that the bacterial photosynthetic unit contains 30 to 50 BChl molecules; these may vary from species to species.

When cells of purple bacteria are disrupted they release a class of subcellular particles containing all the photosynthetic pigments. These pigment-bearing particles can be isolated by differential centrifugation. When examined by electron microscopy after staining, the particles appear like spherical bodies 30 to 100 nm in diameter and are called *chromatophores*.

Table 7.1. *Characteristics of photosynthetic bacteria*

Group	Photosynthetic pigments	e^- donor (substrate for growth)	Growth conditions and other properties	Examples
1. Greens (a) Chlorobiaceae	BChl a plus BChl c, BChl d or BChl e Carotenoids Reaction centre P840 (BChl a)	H_2S $Na_2S_2O_3$ H_2	Light, autotrophic; strict anaerobes; non-motile PS apparatus: *Chlorobium* vesicles and associated membranes (chlorosomes), PSI-type RC	*Chlorobium limicola, Ch. thiosulfatophilum, Prosthecochloris aestuarii*
(b) Chloroflexaceae	Similar to (a)	Organics	Facultative aerobic; filamentous gliding; chlorosomes; PSI-type RC	*Chloroflexus aurantiacus*
2. Purple sulphur bacteria (Chromatiaceae)	BChl a or BChl b Carotenoids Reaction centre P870 or P890	H_2S, $Na_2S_2O_3$ H_2 Organic, e.g. acetate	Autotrophic and heterotrophic in light; strict anaerobes; PS apparatus: chromatophores; PSII-type RC.	*Chromatium D Thiocapsa roseopersicina*
3. Purple non-sulphur bacteria (Rhodospirillaceae)	BChl a or BChl b Carotenoids Reaction centre P870 or P960	Organic, e.g. succinate, malate H_2	Heterotrophic or autotrophic in light and anaerobic; will grow aerobically and heterotrophically in the dark; PSII-type RC	*Rhodospirillum rubrum Rhodopseudomonas viridis, Rhodobacter sphaeroides*
4. Heliobacteria	BChl g Reaction Centre P798	Organics	No chromatophores or chlorosomes; PSI-type RC	*Heliobacterium chlorum H. mobilus*

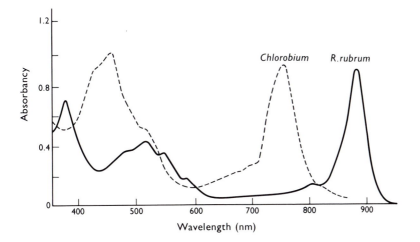

Fig. 7.1   Absorption spectra of green (*Chlorobium*) and purple (*Rhodospirillum rubrum*) photosynthetic bacteria.

Each chromatophore contains several photosynthetic units. The chromato-phores prepared from *Rhodobacter sphaeroides*, for example, consist of approximately 40 reaction centre complexes, 500 light-harvesting (LH) complexes, 1000 carotenoids and 1000 ubiquinone molecules. They are probably derived from the external cytoplasmic membrane by extensive invaginations (infolding) of the membrane.

Duysens was the first to show that the absorption spectrum of bacter-iochlorophyll in purple bacteria is changed reversibly by illumination. The change corresponds mainly to a bleaching (oxidation) of the long-wave absorption band of bacteriochlorophyll: at 960 nm in *Rps. viridis*, at 890 nm in *R. rubrum* and *Chromatium*, and at 870 nm in *Rb. sphaeroides* – these are now called reaction centre BChls.

By treating chromatophores of *Rb. sphaeroides* with detergents and then illuminating them in oxygen, it is possible to destroy the light-harvesting bacteriochlorophyll molecules while still keeping the light-reacting com-ponent intact. The 870 nm absorption band of such specially prepared chromatophores is oxidized reversibly by light. This special light-reacting component in *Rb. sphaeroides* is called P870 and its role is similar to that of P700 in chloroplasts. The equivalent component in *R. rubrum* and in *Chromatium* is designated P890. There is one P870 (or P890) molecule for about 40 bacteriochlorophyll molecules which constitute the photosyn-thetic unit (Fig. 7.2). Oxidized minus reduced difference spectroscopic studies (see Chapter 4) showed the reversible oxidation and reduction of a

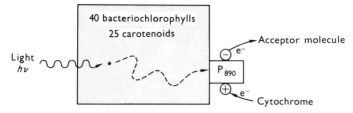

Fig. 7.2　Diagrammatic representation of the photosynthetic unit in bacteria. One photon of light reacts in a unit of 40 bacteriochlorophyll molecules containing one P890 reaction centre.

quinone (ubiquinone) and of a cytochrome simultaneously with the light-induced changes in P870, indicating the participation of these cofactors in photosynthetic electron transport.

The photosynthetic pigments of the green bacteria are located predominantly in specialized structures called *chlorosomes*. Freeze-fracture EM studies of *Chlorobium* and *Chloroflexus* chlorosomes have contributed towards our understanding of the ultrastructure of these organelles. The X-ray crystallographic structure of *Chlorobium* chlorosomes is known also. The chlorosomes are cigar-shaped vesicles, about 100×30 nm in size, bounded by a thin (2–3 nm) single-layered envelope (possibly a lipid monolayer). They are packed with chlorosome cores which consist of rod elements approximately 10 nm in diameter and containing BChls *c* and *d*. The antenna polypeptides are sequestered within the core. The chlorosomes are attached to the inside of the cytoplasmic membrane where the reaction centres are housed, via a proteinaceous *baseplate* containing antenna BChl *a* (Fig. 7.3). The chlorosome pigments constitute the bulk of the very large antenna found in all green bacteria, which may be as high as 1500 bacteriochlorophylls per reaction centre.

Cells of *Heliobacterium* do not show any of the characteristic membrane components seen in other photosynthetic prokaryotes such as chromatophores, chlorosomes or phycobilisomes. The reaction centre is tightly bound to a plasma membrane and contains BChl *g*, which has major *in vivo* absorbances at 788, 576 and 370 nm. The reaction centre chlorophyll, identified by its reversible photobleaching at 798 nm, is designated P798. It has an E_m of $+225$ mV and the 'primary electron acceptor' is 8-hydroxy Chl *a* (BChl 663). The secondary acceptors are iron sulphur centres; the ultimate Fe–S cluster has an E_m of -510 mV measured at pH 10 (the Fe–S centre is not chemically reducible at pH 7). Thus the electron transport properties of *H. chlorum* are those of PSI type and are similar to those of green sulphur

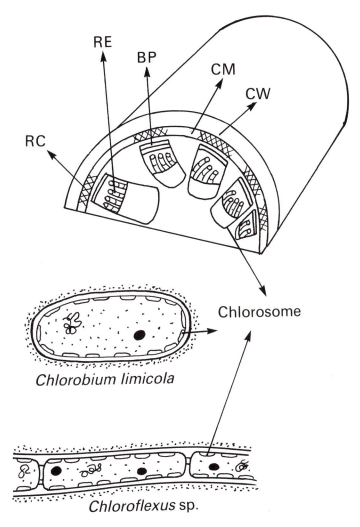

Fig. 7.3   Diagram of two species of green bacteria, and a 'cut out' showing the location of the chlorosomes and some constituents within the chlorosome. BP, baseplate; CM, cytoplasmic membrane; CW, cell wall; RC, reaction centre; RE, rod element.

bacteria (Fig. 7.8). *H. chlorum* has a membrane-bound *c*-type cytochrome which presumably is the electron donor to photo-oxidized P798.

In recent years, due to improvements in the techniques of isolation and crystallization of bacterial reaction centres and the introduction of pico- and femto second laser flash spectrophotometry, most of the events occurring in the light reactions of bacterial photosynthesis have been

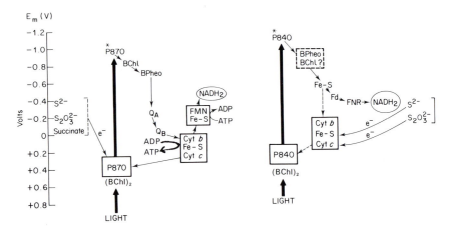

Fig. 7.4 Scheme of electron transport in photosynthetic bacteria. **Left:** purple; **right:** green.

elucidated. Light energy absorbed by bacteriochlorophylls and carotenoids is channelled to a reaction centre containing a few (two or four) specialized BChl molecules. Charge separation occurs across the membrane at these BChl molecules with 100% efficiency (quantum yield = 1), followed by electron transport resulting in the production of ATP, $NADH_2$ or reduced ferredoxin (Fig. 7.4 and §7.3).

7.3 Photochemistry and electron transport

By selective use of different types of detergents, it is possible to isolate pure reaction centres (RC) from chromatophores and study their spectral and structural properties. The most fully characterized reaction centres are those from *Rb. sphaeroides* and *Rps. viridis*. They have molecular mass of approximately 80 kDa and is built up of three polypeptides of molecular mass 21 kDa, 24 kDa and 32 kDa, designated L, M and H subunits respectively. Each reaction centre contains 4 BChl, 2 BPheo, 1 ferrous iron, 2 quinones (one loosely and the other tightly bound), and 1 carotenoid. The composition of reaction centres of other purple bacteria are more or less similar – the major difference being the presence of *c*-type cytochromes in some RC complexes. The RC unit of *Thiocapsa pfennigi*, for instance, contains 1 P960, 4 BChl, 2 BPheo, 1 carotenoid, 2 quinones, 2 cytochromes (*c*-555 and *c*-557) and 5 polypeptides.

The RC unit of *Rps. viridis* was crystallized and its structure determined

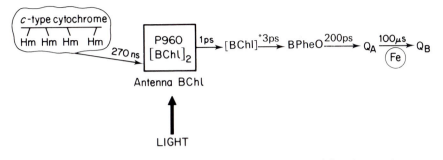

Fig. 7.5 Light-induced electron transfer sequence and time scale (t½) in the reaction centre of *Rhodopseudomonas viridis*. BChl, bacteriochlorophyll; BPheo, bacteriopheophytin; Hm, haem; Q, quinone; *, excited state; ps, picoseconds. (After Deisenhofer, Michel and Huber, 1985.)

at atomic resolution by X-ray diffraction analysis. The crystal unit $(3 \times 7 \times 13 \, nm)$ has molecular mass of 150 kDa and is embedded in a protein moiety made up of four polypeptides (Deisenhofer, Michel and Huber). There are 4 BChl *b*, 2 BPheo *b*, 1 menaquinone (Q_A), 1 ubiquinone and a *c*-type cytochrome (see Plate II). Two of the BChl *b* constitute the primary electron donors, the other two are denoted as accessory BChl. The crystals are photoactive, electron transfer from the energy trap bacteriochlorophyll pair (BChl *b*)$_2$ to the primary acceptor Q_A occurring in 200 picoseconds (Fig. 7.5). This work was soon followed by the determination of the crystal structure of the RC of *Rb. sphaeroides* by Feher and colleagues in California.

Pure RC complexes, completely free of core LH pigments, have not yet been prepared from the Chlorobiaceae. However, photoactive RC structures bound to proteins have been partially purified from *Prosthecochloris aestuarii* and are found to be similar to the PSI in chloroplast. Fully active RC complexes free of core LH pigments have been isolated from *Chloroflexus aurantiacus*, a member of the Chloroflexaceae. They contain 3 BChl *a*, 3 BPheo *a*, and 2 polypeptides per RC, and are structurally similar to the RC complexes isolated from purple bacteria. The photosynthetic electron transport pathway of *Chloroflexus* also is similar to that of purple bacteria.

7.4   Carbon dioxide fixation

Photosynthetic bacteria do not show an Emerson enhancement effect, so it is generally considered that there is only one major photoreaction in photosynthetic bacteria. In cell-free preparations of photosynthetic purple

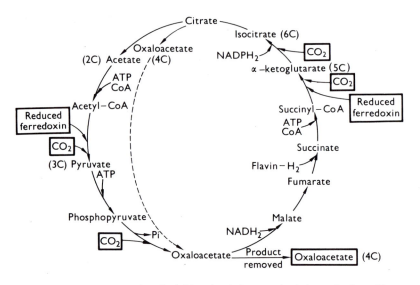

Fig. 7.6   The reductive carboxylic acid cycle of photosynthetic bacteria. Four CO_2 molecules are fixed to form one oxaloacetate molecule.

bacteria, the dominant photosynthetic reaction is cyclic photophosphorylation (production of ATP). Particles have been prepared from green bacteria which are able to catalyze the photoreduction of ferredoxin, which can subsequently be used to reduce NAD to $NADH_2$ via a flavoprotein enzyme. Thus the photosynthetic bacteria can generate both ATP (energy) and $NADH_2$ (reducing power) for the fixation of CO_2 (see Fig. 7.4). The Calvin cycle enzymes are shown to be present in nearly all strains of photosynthetic bacteria studied which thus have the ability to fix CO_2 via this cycle. Evans, Buchanan and Arnon, working in California, demonstrated in 1966 the existence of an alternative route for photosynthetic CO_2 fixation in green bacteria which is mediated by reduced ferredoxin and leads to the synthesis of α-keto acids, e.g. pyruvate and α-ketoglutarate. Their scheme of reductive carboxylic acid cycle in photosynthetic bacteria is shown in Fig. 7.6. In this pathway of photosynthetic CO_2 fixation, the ferredoxin (Fd)-dependent enzymes pyruvate synthase and α-ketoglutarate synthase catalyze the carboxylation of acetyl and succinyl-coenzyme A, respectively, as shown:

(i) Acetyl CoA $+ CO_2 + Fd_{red} \xrightarrow{\text{enzyme}}$ Pyruvate $+$ CoA–Fd_{ox}

(ii) Succinyl CoA $+ CO_2 + Fd_{red} \xrightarrow{\text{enzyme}} \alpha$-ketoglutarate $+$ CoA $+ Fd_{ox}$

The overall cycle involves the fixation of four molecules of CO_2. It is probable that in these bacteria both the reductive pentose phosphate (Calvin) pathway and the reductive carboxylic acid cycle are operating in photosynthetic CO_2 fixation; the former being the major route for CO_2 assimilation in the 'purples', and the latter the main route in the strictly anaerobic green sulphur bacteria.

7.5   Light energy conversion by halobacteria

A unique type of light energy conversion for ATP synthesis is shown by certain halobacteria typified by *Halobacterium halobium*. These organisms normally grow in aerobic, extremely saline environments (3.5 to 5 M NaCl). However, when cultured at low O_2 concentrations and high light intensity *H. halobium* cells form patches of a purple membrane on their surface which contain two photoactive proteins, *bacteriorhodopsin* and *halorhodopsin* (retinal protein). Bacteriorhodopsin can use the visible spectrum of light energy to generate a proton gradient from the interior of the cell to the exterior. This transmembrane proton gradient is coupled via an ATPase for the synthesis of ATP by the bacterium. Halobacteria lack antenna pigments and the carotenoids of *H. halobium* cannot transfer light energy to bacteriorhodopsin. So, the efficiency of energy conversion is probably lower than that in chlorophyll-mediated photosynthesis. Halorhodopsin absorbs (green) photons and can function as a Cl^- pump directed inwardly from the membrane.

7.6   Ecological significance of phototrophic bacteria

Under anaerobic (O_2-free) conditions, organic matter is fermented by various micro-organisms (the chemosynthetic anaerobes) which gain their energy by a substrate-linked phosphorylation. Various metabolic end-products such as CO_2, H_2, ethanol and simple fatty acids are formed during this process. Such compounds would accumulate if they were not removed as nutrients by other types of microbes which are unable to use O_2 as the ultimate electron acceptor in their respiratory processes. The sulphate- and nitrate-reducing bacteria are able to consume part of the end-products of fermentation of the chemosynthetic anaerobes. The phototrophic bacteria (green and purple photosynthetic bacteria) derive their energy from light and are able to metabolize most of the end-products of anaerobic fermentation such as alcohols, acids and hydrogen, as well as the end-products of

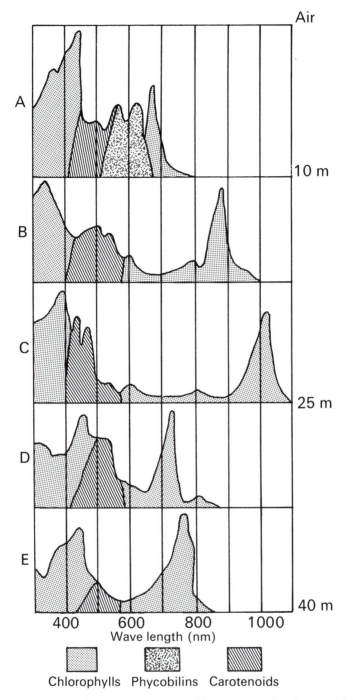

Fig. 7.7 Cellular absorption spectra of five representative phototrophic bacteria. **A:** Cyanobacterium, Chl *a* and phycobiliproteins; **B:** purple bacterium; BChl *a*; **C:** purple bacterium: BChl *b*; **D:** green bacterium BChl *e* and *a*, **E:** green bacterium BChl *c* and *a*. The heights in the Y-axis show the approximate depths the phototrophs can occupy in a water column. (After Stanier *et al.*, 1989.)

sulphate and nitrate respiration such as H_2S and N_2. Thus the cell materials synthesized by the green and purple photosynthetic bacteria are future substrates for the chemosynthetic anaerobes which again produce the nutrients for the phototrophic bacteria. Thus these two types of bacteria growing in an O_2-free environment can exist together.

Phototrophic bacteria occur ubiquitously. Cyanobacteria are found growing in deserts, volcanic regions, thermal hot springs, polar ice caps and even on top of Mount Fuji! More than half the total primary productivity in the open ocean is due to prokaryotic phytoplankton. They play a major part in the cycling of C, N_2, O_2, S, P and H_2.

The optimal light for photosynthesis for cyanobacteria is between 20 and 250 μmol m^{-2}s^{-1} and 5–10 μmol m^{-2}s^{-1} for purple phototrophs. Higher ambient light intensities promote the synthesis of increased levels of photoprotective pigments (carotenoids) or light-avoidance strategies such as moving down the water column. In aquatic systems, shorter wavelengths of visible spectrum (500 to 700 nm) are absorbed first by the water on the top layer and also by the oxygenic photosynthesizers (cyanobacteria and microalgae) growing on the water–air interface. Longer wavelengths (700 to 1100 nm) are absorbed in the deeper water column. Purple and green bacterial colonies abound in a stagnant layer of water at depths of 10 to 30 metres that is essentially O_2 free. Although most purple bacteria have pigments that absorb in the near infra-red (BChl *a*, 800 to 850 nm), some species can absorb wavelengths up to the infra-red (BChl *b*, 1015 nm). Green sulphur bacteria have adapted to very low light intensities by increasing the amount of antenna pigments (mainly carotenoids) relative to the reaction centre bacteriochlorophylls. The absorption characteristics of the pigments in phototrophic bacteria are shown in Fig. 7.7.

7.7   A comparison of plant and bacterial electron transport

(a) *Common features*   There are many basic processes which are common to all three types (purple and green bacterial and oxygenic) of photosynthesis. The first step in photosynthesis, photon capture, is performed by light-harvesting (LH) antennae which consist of an array of photosynthetic pigments conjugated to proteins. The differences in LH properties among the photosynthetic cells lie in the composition of the accessory pigments of the antenna complex. In the green sulphur bacteria the LH components are located in non-membranous, particulate structures (chlorosomes, see Fig. 7.3), whereas in the purple bacteria the LH and RC complexes are held together within a membrane. The phycobilisomes of red algae and cyanobacteria are located on the outer surface of the photosynthetic membrane. The light-harvesting complexes in the PSI and PSII of chloroplasts (of algae and higher plants) are attached to the thylakoid membrane. Light energy absorbed by the antenna complex in all photosynthesizers is delivered to the RC by resonance transfer with an efficiency approaching 100%. This high photosynthetic efficiency is achieved by regulating the synthesis of LHC components in response to the incident light quality and intensity and to the nutrient availability.

The conversion of light energy to chemical potential occurs in the reaction centres located in lipid bilayer membranes. The reaction centres are membrane-spanning proteins to which are bound the primary donor P ($[Chl]_2$ in PSII, $[BChl]_2$ in bacteria), and the cofactors that mediate electron transfer from P to a 'tertiary' acceptor (quinones or Fe–S proteins) (Fig. 7.8). The primary photoreaction in all systems is a light-induced electron transfer from P to a primary acceptor. The relatively low energies of the excited donor P* at the RC helps to minimize the wasteful decay of the excited states via fluorescence, with the result that the quantum yields of electron transfer reactions are high at the RC complex. Back reactions are prevented by the ultra fast rate of electron conduction from P* through a series of donor–acceptor species which are stabilized for increasingly longer distances.

There is a considerable drop in redox potential between the primary and the final stable electron acceptor; the actual range of redox potential between the two acceptors is different for various reaction centres. In the green sulphur bacteria (primary acceptor not yet precisely identified) and in photosystem I of plants, the potential falls from -1.0 V (A_0 in PSI) to about $-0.730/-0.550$ V (reduced Fe–S centres). Those reaction centres where the final stable electron acceptor is a reduced [Fe–S] protein and which function at a low (electronegative) redox potential are known as the PSI type. In purple bacteria and PSII, the potential drops from -0.60 V (Pheo* or BPheo*) to $+0.08$ V$/0.0$ V (Q_Bs). Those RCs which function at a more electropositive (high) potential and where a quinone is the final stable acceptor are referred to as the PSII type.

(b) *Reaction centres of PSII and purple bacteria*   The RCs of purple photosynthetic bacteria and of photosystem II of plants show many similarities [Fig. 7.8). In both RCs the primary electron donor is a special chlorophyll pair [Chl]$_2$ and the primary acceptor is a pheophytyn. Charge recombination in the primary donor–acceptor radical pair [Chl$^+$ Pheo$^-$] creates a triplet state in both instances with similar EPR spectra. Electrons are transferred from reduced pheophytin (Pheo$^-$) to the primary quinone Q$_A$ (plastoquinone in PSII menaquinone in bacterial RC) which is a one-electron acceptor bound to a protein. The secondary quinone Q$_B$ (plastoquinone in PSII and ubiquinone in bacterial RC) is a two-electron carrier which is tightly bound to protein only when it is in the semiquinone (one-electron) form. After double reduction and protonation, Q$_B$ is converted to a quinol which dissociates from the RC; its site in the protein is then occupied by a different Q$_B$ molecule from the quinone pool. The quinones in both RCs are similar with respect to their kinetic properties and their interaction with the ferrous iron atom. Atrazine-type herbicides inhibit electron transfer in both RCs; studies with mutants showed that these types of inhibitors act by displacing Q$_B$ from its binding site in the RC protein.

X-ray crystallographic analyses of the isolated reaction centres from the purple sulphur bacteria, *Rhodopseudomonas viridis* and *Rhodobacter sphaeroides* have revealed the presence of three membrane-bound polypeptides referred to as the L, M and H subunits. The amino acid sequences of the bacterial L and M subunits and of D$_1$ and D$_2$ polypeptides from spinach PSII show homology (i.e. certain identical amino acids occur in the same positions in the four polypeptides L, M, D$_1$ and D$_2$), suggesting a possible common ancestor for the four polypeptides. Homology in protein or nucleic acid sequences isolated from different organisms is one of the criteria used by biochemists to trace the evolutionary relationship between the organisms. The X-ray data of bacterial RC indicated a symmetric arrangement of the pigments and the L and M subunits in the membrane; the subunits provided binding sites for the primary donor (P) and the quinone–iron complex. Elegant studies by site-directed mutagenesis, i.e. genetic alteration of specific amino acids, and by the use of selective inhibitors, showed that in both L and D$_1$ proteins the amino acids Phe and Ser, which provide binding sites for Q$_B$, occur in identical positions in the protein chains. A trytophan residue, presumed to be involved in the binding of Q$_A$, occurs in identical sites in M and D$_2$ proteins. The similarities in the location of the key functional amino acid residues, the nature of their protein folding, and the number of membrane-spanning helices between L

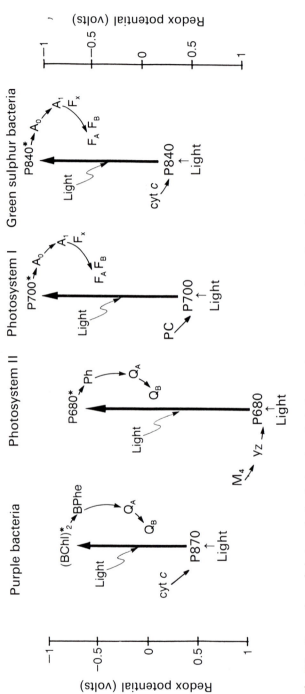

Fig. 7.8 Comparison of photosynthetic electron transfer in the reaction centre complexes of purple and green bacteria and in chloroplast PSI and PSII. (Redrawn from Nitschke and Rutherford, 1991.)

and M in bacterial RC and D_1 and D_2 in PSII led to the suggestion that D_1 and D_2 constitute the PSII reaction centre core. Support for this postulate only came later with the isolation of PSII RC core complex from spinach with D_1 and D_2 proteins intact. Although this core complex showed light-induced electron transfer from an artificial donor to pheophytin (in the core), the plastoquinones were lost during the isolation of the RC core complex.

(c) *Differences*   Are there significant differences between the purple bacterial and PSII reaction centre complexes? Certainly yes. The most important difference is the absence of a water-oxidizing cluster in the purple bacteria. We have already noted the absence of Q_A in the PSII RC core, indicating that the quinone is less tightly bound in the PSII RC. The isolated PSII core also contains cyt b559 and the two chlorophyll–protein complexes CP43 and CP47. Bacterial (wild-type) RCs are insensitive to phenolic and urea-based (e.g. DCMU) herbicides which inhibit electron transport in PSII. Bicarbonate has been suggested as a fifth ligand for the ferrous iron which is bound to the D_1–D_2 dimer, whereas the fifth iron ligand in the bacterial RC may be a glutamic residue in the M subunit. Amino acid sequence comparisons provide no clue to the binding sites of the accessory chlorophylls in D_1 and D_2; maybe these chlorophylls are bound in a different way in PSII RC. There are two histidine residues (at position 118) in D_1 and D_2 but not in L and M. Are these histidines involved in the binding of Mn (of the water-oxidizing cluster) or of chloride ions? These questions can be answered convincingly only after crystallization of the PSII RC core and the determination of its structure by X-ray analysis.

(d) *Green sulphur bacteria and PSI of plants*   The main characteristics of these RCs are their ability to photoreduce bound Fe–S centres that have very low redox potentials (< -0.5 V). The Fe–S centres in green bacteria are at least 0.15 V more reducing than in plant PSI. The final product of photosynthetic electron transport in the green sulphur bacteria is $NADH_2$, as is $NADPH_2$ in the PSI of plants. Compared to the data available for PSII and the purple bacterial RC, knowledge of the PSI RC is still insufficient.

The green sulphur bacteria *Chloroflexaceae* are similar to the purple sulphur bacteria in RC composition and structure. Looking beyond the reaction centres, in plants as well as bacterial photosystems, a quinone–cyt b–cyt $c(f)$ assembly functions as a gate between a two-electron carrier and a one-electron carrier via a 'Q cycle' in which protons are transported across the membrane to synthesize ATP (§8.10).

7.8   Evolution of photosynthesis

Early attempts to trace evolutionary relationships between organisms were based on morphological and metabolic traits and on comparative amino acid sequence analysis of certain proteins (e.g. ferredoxin and cytochrome c) involved in the energy metabolism of the organisms. It was believed that the evolution was the result of gene mutations and gene fusions. Subsequently, 16s ribosomal RNA sequence analysis was introduced as a tool for the construction of phylogenetic relationships. The conclusions derived from 16s RNA sequences were at times in conflict with the evolutionary relationships drawn from amino acid sequence studies; the idea of lateral gene transfer between photosynthetic bacteria was put forward to explain some of the anomalies. Data gathered from biochemical and biophysical studies on the structure and function of the reaction centres of photosynthetic organisms have provided an additional basis for their classification.

Fossil evidence clearly shows the existence of organisms similar to modern-day cyanobacteria on the earth 3.5×10^9 years ago, which suggests that oxygenic photosynthesis was established on the earth about that time. How did the cyanobacteria originate? From the similarities in properties of the purple bacterial RC and that of PSII RC, and from the 16s RNA sequence analysis of purple and green sulphur bacteria, it has been suggested that the cyanobacteria would have originated from a genomic fusion between a bacterium having a purple bacterial PSII-type RC and a bacterium having a PSI-type RC. We have no clue as to when and how a water-oxidizing machinery was incorporated into this gene fusion product.

Evolutionary schemes for photosynthetic reaction centres have been put forward by Nitschke and Rutherford and by Blankenship. Both schemes propose a common ancestor-type RC from which a prototype of the PSII-type RC, purple bacterial RC, and *Chloroflexus* RC evolved on the one hand, and the PSI-type RC, green sulphur bacterial RC, and the heliobacterial RC evolved on the other hand by separate divergent pathways. Blankenship further proposes that the cyanobacteria and chloroplasts may have originated by gene duplication of the PSI and PSII-type reaction centres followed by fusion of the duplicated gene products. These classifications do

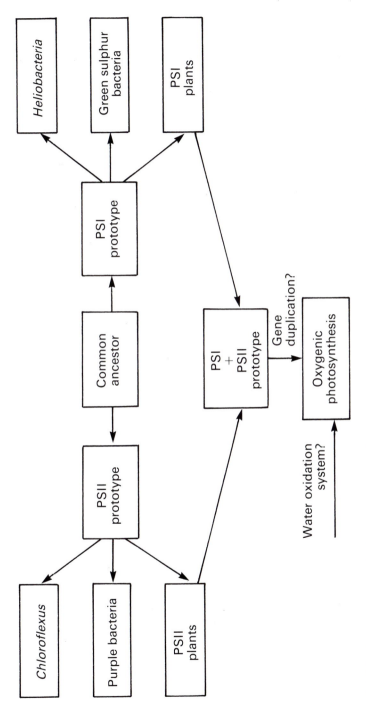

Fig. 7.9 Proposed scheme for the evolution of photosynthetic reaction centres. (After Nitschke and Rutherford, 1991, and Blankenship, 1992.)

not necessarily take into account other metabolic characteristics of the organisms.

A better picture of the phylogenic tree could be drawn in the future after the determination of the three-dimensional structure of all types of reaction centres. Meanwhile, the scheme shown in Fig. 7.9 takes into consideration current thoughts on the evolution of photosynthetic function.

8

Research in photosynthesis

The main areas of current research into the broad field of photosynthesis are summarized below. Many of these are interrelated and share research methodologies.

8.1 Phytochromes

Phytochrome is a pigment present in all plants and which is essential for the regulation of many stages of plant growth and development. However, unlike the photosynthetic pigments, phytochrome is not involved in light energy transfer and storage. In higher plants phytochrome regulates molecular and physiological processes such as gene expression, seed germination, chloroplast development, stem elongation, leaf expression, flowering etc. Many of these phytochrome-mediated effects require only brief exposure of the plant material to dim *red light*. In the now-classic experiments on the effect of light on lettuce seed germination, it was observed that irradiation of the seeds with red light (650–700 nm) promoted germination; a dose of far-red (710–740 nm), given immediately after the red, inhibited germination. To explain these observations it was postulated, and later (in 1959) verified, that the pigment which causes these changes must itself be photoreversible, i.e. its absorption properties should undergo reversible changes after red and far-red light treatment.

In plants kept in darkness, phytochrome is present as a blue pigment which absorbs red light, this form is referred to as Pr. Exposure to red light converts the blue phytochrome to a bluish-green molecule which absorbs far-red light; this form is referred to as Pfr (Fig. 8.1).

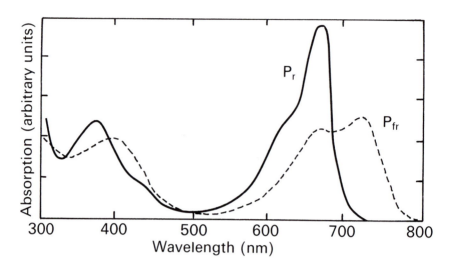

Fig. 8.1 Absorption spectra of the two main forms of phytochromes.

Structurally, phytochrome consists of a tetrapyrrole chromophore cova-lently bound to a protein. The apparent mass of the whole molecule is approximately 124 kDa. Light absorption by the chromophore causes isomeric shifts of the tetrapyrrole double bonds and possibly leads to conformational changes in the protein resulting in the interconversions between the Pr and Pfr forms. In all dicotyledonous plants the Pfr form reverts spontaneously to the Pr form in the dark.

Comparison of the action spectra (measurement of seed germination rate, stem elongation, etc.) of phytochromes with their absorption spectra has shown that most of the physiological responses in isolated organelles (usually short-term effects) and in whole plants are elicited by Pfr. Phytochromes are believed to play a role in circadian rhythms (cyclic variations of physiological activities on a daily basis). These roles are illustrated in Fig. 8.2.

Phytochrome is synthesized in the dark in the Pr state; the rate of synthesis is presumably controlled by light. The overall content of phytochromes in plants is very low ($<0.2\%$) and the molecule is extremely susceptible to proteolytic breakdown during isolation.

The mechanism of action of phytochrome is not yet fully understood. Changes in membrane permeability, activity of membrane-bound enzymes, transcription or translation of genetic information in chloroplasts etc. are some of the effects suggested to be phytochrome-controlled. It is also

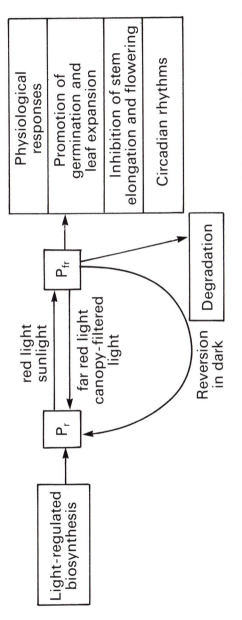

Fig. 8.2 Diagram illustrating the synthesis, photoconversions and physiological effects of phytochromes in cells.

postulated that Ca ions may act as a messenger to trigger phytochrome activity. These postulates are being investigated by molecular geneticists and photobiologists.

8.2   Protoplasts and cells

Protoplasts are basically cells without their cell walls (which have been removed by enzymatic digestion). Interferring substances, such as phenolics, present in cell walls or vascular tissues of leaves are removed during enzymatic digestion, and hence chloroplasts prepared from protoplasts show a high degree of photosynthetic activity compared to the chloroplasts isolated by mechanical disruption of leaves. The refinement of protoplast isolation techniques has extended the range of plants from which active chloroplasts may be isolated. Thus, intact chloroplasts can now be isolated from C_3 grasses (wheat and barley), C_3 dicot plants (tobacco, sunflower) as well as many C_4 species (maize, sugar cane). In C_4 species, controlled enzymic digestion of leaves results in the preferential release of mesophyll protoplasts, leaving the bundle-sheath strands intact. Protoplasts have proved very valuable in the study of inter-and intracellular compartmentation of photosynthetic metabolite activities in C_3, C_4 and CAM species of plants. Individual cells (with intact cell walls) can now also be isolated from the leaf in order to study the more complex interactions in the movement of substrates, ions, gases, etc. The techniques used vary greatly depending on the species and variety of plant being studied. Nowadays, protoplasts are increasingly used for the creation of transgenic plants (§8.4).

8.3   Origin and development of chloroplasts

Did the chloroplasts of algae and higher plants (eukaryotes) evolve from the prokaryotic blue-green algae (cyanobacteria), as proposed in the endosymbiotic theory of evolution of eukaryotes? Are the *cyanelles* (symbiotic algae, see Fig. 3.4, p. 36) which are found in some obligate photoautotrophs such as the *Cyanophora* sp. (a flagellated alga containing a blue-green algal symbiont) modern representatives of evolutionary links between blue-green algae and chloroplasts? Are the prochlorophytes with Chls $a + b$ (e.g. *Prochloron didemnii*) a connecting link between chlorophytes and primitive algae or bacteria? Are the cyanobacteria really the first group of oxygenic photosynthesizers or were there other micro-organisms with photosensitive pigments capable of water photolysis prior to the evolution of cyanobacteria which we know from fossil records existed in the earth about 3.5×10^9 years ago? What criteria should be followed for evolutionary studies: morphology and metabolic functions, homology in protein and DNA sequences, or homology in 16s RNA sequences? These are some of the questions posed by evolutionists (see also §7.7).

A large body of sequence data has accumulated for the chloroplast-located gene *rbc* L which codes the large subunit of RuBisCO. The data cover all types of seed plants and provide opportunities for systematic biologists to construct evolutionary 'trees'. Comparative analysis of available plant and algal chloroplast gene sequences and identification of loci of nucleotide changes (substitution, deletion and addition) in the genetic map have revealed enough information on the phylogenetic origins of plant species. These data, and the computer-assisted programs, should enable biostatisticians to draw up evolutionary schemes for the plants and algae.

Chloroplast development

In angiosperms (flowering plants), chloroplast biogenesis is usually studied by germinating seeds in total darkness and then exposing the seedlings to light. The changes occurring during illumination are followed by examination of the ultrastructure of the plastids by electron microscopy, the assay of photosynthetic electron transport and CO_2 fixation rates, and determination of the chromophore-containing components, e.g. cytochromes and ferredoxins, by spectroscopy. Such studies have shown that chloroplasts develop from relatively undifferentiated structures known as *proplastids* via simple colourless organelles termed *etioplasts*.

The most important change occurring during illumination (and greening) is the reduction of protochlorophyllide to chlorophyllide (chlorophyll without the phytyl side chain). The response of leaves to light involves changes in the expression of specific genes at the RNA level – RNAs complementary to many segments of the plastid genome (see below) are expressed more abundantly in light. Substantial increases in the content of carotenoids, cytochromes, ferredoxin, plastocyanin, ferredoxin–NADP reductase, etc. have been measured during the greening process. Photosystem I activity is expressed at an earlier stage than PSII activity. The enzymes of the CO_2 fixation pathway are present in the etioplasts, albeit at very low concentrations compared to mature chloroplasts. In higher plants the responses to light are thought to be mediated by three photoreceptors: protochlorophyllide, phytochrome (§8.1) and a blue light receptor. The blue light receptor has not yet been isolated, although both flavoproteins and carotenoids have been proposed as likely molecules.

8.4   Chloroplast genetics; expression and regulation of genes; transgenic plants

The presence of DNA and ribosomes in chloroplasts was first demonstrated in 1962. Our present knowledge of the organization of genetic information in chloroplasts is based on the application of molecular cloning and nucleotide sequencing technologies. Restriction mapping and cloning confirmed the double-stranded circular nature of the chloroplast DNA (plastome). The size of the plastome varies between 120 and 160 kbp (kilobase pair) in land plants; the average molecular mass of the plastome is about 1×10^8 kDa.

To date, the entire genome sequence has been determined for the cloned chloroplast DNAs of three plant species, viz. tobacco, liverwort and rice (Fig. 8.3), and the sequences for defined segments of many other chloroplast genomes have been completed. These studies indicate that the chloroplasts probably code for all of their tRNA (transfer RNA) and rRNA (ribosomal RNA) molecules and many of the proteins in the photosynthetic electron transport chain and in the ATP synthase. The remainder of the electron transport chain proteins and some in the ATP synthase complex are encoded by the nuclear genome, synthesized in the cytoplasm, and imported into the chloroplasts (§8.5).

Genetic engineering of the chloroplast genome (i.e. deletions of segments of the gene and/or introduction of foreign DNA into the genome) is now possible and has become a useful tool in basic and applied research (e.g. creation of transgenic plants). As an illustration, the unicellular green alga *Chlamydomonas reinhardtii* has a single chloroplast containing 5 to 80 copies of the 200-kbp circular DNA molecule. The alga can be grown autotrophically in light or heterotrophically in the dark (when the growth medium includes a reduced C source such as acetate). Numerous chloroplast and nuclear gene mutants of *Chlamydomonas* deficient in PSII, PSI, b_6f complex, ATP synthase or LHC have been isolated from the heterotrophically grown alga. Subsequently, foreign plasmid DNA containing the deletion segment, with site-specific alterations in nucleotides, can be introduced and integrated into the algal genome using the biolistic particle transformation system (see below). The proteins expressed from these transformed genome, with site-specific mutations, provide valuable information as to their function and topology in the membrane and also as to the co-ordinated synthesis of chloroplast polypeptides.

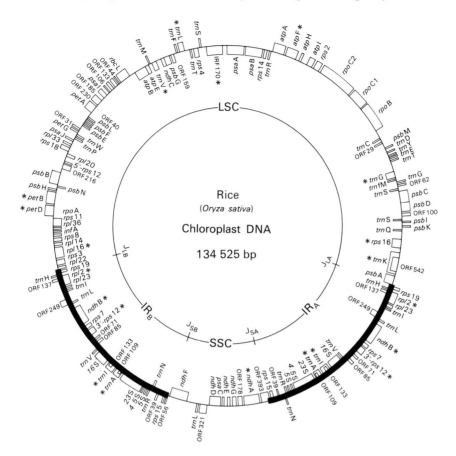

Fig. 8.3   Genetic map of the double-stranded circular *Oryza sativa* (rice) chloroplast genome drawn to scale. Genes shown on the outside of the circle are encoded on the A strand and transcribed counter-clockwise. Genes on the inside are encoded on the B strand and transcribed clockwise. Genes coding for major chloroplast components are: *psa*, PSI proteins; *psb*, PSII proteins, *pet*, photosynthetic electron transfer proteins; *atp*, ATP synthase subunits; and *rbc* L, RuBisCO large subunit. ORF, open reading frame. For details of other genes refer to Hiratsuka *et al.* (1989). *Molecular and General Genetics*, **217**, 185–194. (Courtesy: M. Sugiura, Nagoya University, Nagoya.)

The genetic maps of two cyanobacterial species, *Synechocystis* and *Anabaena*, are nearing completion. Several mutants of *Synechocystis*, lacking specific photosynthetic components, have been isolated; these mutants are used to study the function of specific residues in the protein components (such as Y_Z in D_1 and Y_D in D_2) of PSII. *Anabaena* mutants are being investigated for strains with better efficiencies in H_2 production and N_2 fixation.

Transgenic plants

In higher plants, exogenous DNA can be introduced into somatic cells either by direct gene transfer or via the plasmid vector of the plant pathogen *Agrobacterium*. The transferred DNA (T-DNA) is referred to as a transfer gene and the progeny of the host as a *transgenic plant*.

Although the first transgenic plants were created in 1983, the technique is still routinely applicable only in a few model species such as *Arabidopsis*, tobacco and petunia. The main difficulty appears to be in directing the gene transfer to competent plant cells that are amenable to regeneration.

Agrobacterium is a soil bacterium that infects wounded plant cells producing tumourous growth at the site of infection. The two species used for production of transgenic plants are *A. tumefaciens*, which induces crown galls in the host, and *A. rhizogenes*, which induces hairy root diseases. Both species carry a large plasmid, tumour-inducing Ti or root-inducing Tr, on which the T-DNA involved in gene transfer and a virulence *vir* region are located. The function of the *vir* region is to attach the bacterium to the plant cell. Transformation is effected by the transfer of a piece of DNA, the T-DNA, from the bacterium to the plant. The T-DNA gets integrated into the plant genome. Selection is achieved by culturing the cells, after transformation, in specific antibiotics such as kanomycin, hydromycin, ampicellin etc.; the plasmids carry segments conferring antibiotic resistance and those cells containing the incorporated Ti-plasmids will survive and grow. Transformed cells can also be identified visually by the use of specific molecules known as reporters. Transformation of genes into plant cells is not the sole criterion of creation of transgenic plants; the cells should be able to regenerate with the acquired genes for several generations.

Transgenic plants offer scope for plant breeders for production of improved crops conferring traits such as herbicide resistance, resistance to insects, viruses, frost etc. However, thus far, *Agrobacterium*-mediated gene transfer has been accomplished only in dicotyledonous plants; the technique has not been successful with cereals and many monocotyledonous crop plants.

Transgenic plants are useful for the study of gene expression and the communication between different genomes in the eukaryotic cell. The expression of many chloroplast genes is regulated by light and presumably is under phytochrome control; how phytochromes control gene expression is not known. Preliminary studies of transcriptional activity in chloroplasts indicate that gene expression is regulated both at the transcriptional and post-transcriptional levels and may involve chloroplast RNA-binding proteins. Although much of the pioneering work on plant genetics was carried out on tobacco (*Nicotiana tabacum*), most of the current investigations are performed on the weed *Arabidopsis* (e.g. thale cress *Ar. thaliana*). The *Arabidopsis* plant is relatively smaller in size, has a short generation time, small genome size and has several identified genetic markers; this weed has become the plant geneticists 'fruit fly'.

What are the prospects for breeding transgenic crop plants? As mentioned earlier, so far attempts to incorporate the *Agrobacterium* plasmid vector directly into cells of monocots and generate transgenic plants have failed. In the last 10 years, several reports describing techniques to transfer DNA into plant protoplasts have appeared. Some of these methods with potential for creation of transgenic cereals are as follows. (1) *Direct transfer* of DNA to protoplasts in the presence of polyethylene glycol (25% w/v PEG) and calcium ions. Success has been reported in the production of fertile transgenic rice. (2) *Electroporation.* High-voltage electric pulses are applied either directly or indirectly to a solution containing plasmid DNA and proplasts, sometimes in the presence of PEG. Electroporation permeabilizes biomembranes. The technique has been applied successfully to rice, wheat and sorghum, and is useful in the study of expression and inheritance of genes introduced into graminaceous monocots. (3) *Microlaser.* Holes are burnt in cell walls and membranes with a microlaser beam and the perforated cells are then incubated in DNA solution. (4) *Microinjection*, in which naked DNA is injected mechanically into the protoplast. (5) *Biolistics* or particle gun (applicable to cells and tissues). Plasmid carrying the chimeric (foreign) gene is absorbed to the surface of microscopic tungsten particles (microprojectiles). The microprojectiles are accelerated to high velocities and shot into intact cells or plant tissue. The particles penetrate the cell walls and the chimeric DNA gets incorporated. Success has been reported in transferring chimeric genes into intact maize cell walls. The technique is capable of delivering DNA into embryogenic tissues.

8.5   Transport and assembly of cytoplasmically assembled polypeptides into the chloroplast membranes; exchange of ions and metabolites through the chloroplast envelope

The majority of the protein components of chloroplasts are encoded in the DNA and are synthesized on cytoplasmic ribosomes outside the organelle and subsequently imported into the chloroplasts. *In-vitro* studies using the protein-synthesizing machinery of a wheat-germ extract coupled to poly-adenyl messenger RNAs (PolyA$^+$mRNA) isolated from algal or higher plants have demonstrated the transfer of a number of nuclearly coded polypeptides into the chloroplast; examples are ferredoxin, plastocyanin, ferredoxin–NADP reductase, polypeptides associated with antenna pigment complexes, the small subunit of ribulose-bisphosphate carboxylase, and fructose 1,6-bisphosphatase. These proteins are assembled on the cytoplasmic ribosomes as a precursor of much higher molecular weight, referred to as the transit peptide, with sequence extension, at the aminoterminal. The precursor of plastocyanin (molecular mass = 10.5 kDa), for example, is assembled with a transit peptide of molecular mass 15 kDa. The precursors are bound to the outer chloroplast membrane, translocated through the chloroplast envelope, processed to the mature polypeptide (mainly by proteolytic cleavage of the transit peptide), and then either assembled as membrane constituents or incorporated into the stroma. The import of precursors into the chloroplast is an energy-dependent process and consumes ATP synthesized by photophosphorylation. The requirements for binding of precursors to the envelope membrane, the nature of the receptors, and the mechanisms regulating the translocation and assembly of polypeptides are still being actively investigated.

Recent studies show there are two classes of precursor proteins based on their thylakoid assembly requirements. The integral membrane proteins (e.g. Chl *a/b* protein of LHC II) must cross the plastid envelope membranes, the aqueous envelope periplasm, the stromal compartment and then integrate into the thylakoid lipid bilayer. These precursor proteins contain a hydrophobic N-terminal extension called the *transit sequence* that contains topogenic information for targeting the precursor to the chloroplast and localizing the matured protein on the thylakoid. Lumenal proteins (e.g. plastocyanin) must cross all the plastid compartments and in addition must traverse the thylakoid bilayer membrane. The plastocyanin precursor contains two transit sequences as chloroplast-targeting components. A chloroplast-targeting domain (at the N-terminal) directs the preprotein towards and into the chloroplasts; this domain is cleft by a stromal protease. A second, strongly hydrophobic, domain at the C-terminal directs the partially cleft precursor to the lumen where it is split off by a thylakoid protease.

Proteinaceous components exposed at the chloroplast surface may be involved in the import and binding of precursors.

At present there is no evidence for cytosolic factors being involved in protein translocation other than, maybe, as chaperones. The molecular *chaperones*, incidentally, are a class of proteins found in prokaryotic and eukaryotic cells which play roles in transport, folding, and assembly of certain other proteins but are not themselves components of the final oligomeric structures. The 70-kDa *heat-shock proteins* (Hsp 70), which are synthesized in cells in response to exposure to elevated temperature (heat shock) and other 'stresses', are a typical group of chaperones. Hsp 70s have been detected in outer envelope membranes and stroma of chloroplasts.

Exchange of ions and metabolites

Electron micrographs of chloroplasts isolated from higher plants show that they are bound by two distinct membranes which are together called the chloroplast envelope (Fig. 3.2b). These envelopes are the site of synthesis of a number of membrane constituents and are involved in the transport of many metabolites, and have thus attracted considerable attention. The envelopes can be separated from intact chloroplasts (or protoplasts) by gentle osmotic shock followed by centrifugation in a discontinuous sucrose gradient – the ligher envelopes sediment above the heavier thylakoids in the gradient. Techniques have also been developed to separate the inner from the outer membranes of the isolated envelopes. Purified envelopes are yellow in colour (due to carotenoids) and are free of chlorophyll, cytochrome *b* and oxidoreductase enzymes. The envelope has been shown to be the site of synthesis and assembly of membrane galactolipids, α-tocopherol and carotenoids. The polypeptide composition of the envelope is different from that of the thylakoids and of the stroma. Polyacrylamide gel electrophoresis of envelope proteins released by detergent treatment shows 75 different polypeptides; some of these are involved in regulating the exchange of metabolites through the envelope and others in maintaining the structure of the membrane.

The chloroplast envelope functions both as a barrier separating the stroma from the cytosol and as a bridge allowing the transport of specific metabolites between these two compartments. The permeability of various molecules through the chloroplast envelope has been studied (a) by feeding $^{14}CO_2$ to leaves and following the distribution of radioactively labelled metabolites in the chloroplast and cytosol, and (b) by adding metabolites to a suspension of *intact* chloroplasts and measuring the kinetics of specific photosynthetic reactions. Such studies have shown that the outer envelope membrane acts as a molecular sieve and is permeable to all molecules up to a molecular mass of 10 kDa. The inner membrane (osmotic barrier) is almost impermeable to charged species such as ferredoxin, NAD(P), NAD(P)H$_2$, acetyl CoA, pyrophosphate, many sugar phosphates, small cations (Na$^+$, Mg^{2+}), and to sucrose and sorbitol. Uncharged molecules such as CO$_2$, O$_2$, acetic acid, pyruvic acid, glycerol, etc. can diffuse through the inner membrane. Passage of charged species such as Pi and triose phosphates, ATP, and dicarboxylate anions (malate, aspartate, oxaloacetate, etc.) through the inner membrane occurs only in the presence of specific carriers present in the envelope called *translocators*. Of these translocators, the proteinaceous factor (phosphate or C$_3$ translocator), which allows the transport of triose phosphates from the stroma to the cytosol in exchange for the import of Pi into the stroma, has been studied in most detail. The importance of this counter-exchange in sucrose synthesis has been stressed earlier (§6.5). The chloroplast envelope also mediates in the export of fatty acids synthesized in the chloroplast to other organelles (for lipid synthesis).

What is the exact nature of the polar lipid interactions with plastid membrane proteins? How are the membrane components which are synthesized in the envelope subsequently transferred and assembled in the thylakoids? Where do the polyunsaturated fatty acids in the chloroplasts originate? What are the specific functions of various envelope proteins? These are some of the questions for which answers are still needed.

8.6   Chloroplast structure

Significant advances have been made towards the understanding of the molecular structure of the thylakoid membrane, i.e. the proteins and lipids and their orientation, the polypeptides associated with light-harvesting pigment–protein complexes and their distribution in the core and peripheral parts of the antennae, etc. Membrane complexes can be isolated from higher plant chloroplasts, and their photochemical and biochemical properties studied *in vitro*. There is strong evidence of a lateral heterogeneity in the distribution of these complexes between stacked and unstacked regions of the thylakoid membrane (Fig. 8.4). Photosystems I and II are now found to exist as two structurally (and even functionally) independent types designated as PSI$_a$, and PSI$_\beta$, and PSII$_a$ and PSII$_\beta$. PSII$_a$ is located in the grana membranes and PSII$_\beta$ is exclusively located in the stroma-exposed thylakoids. The light-harvesting capacity of PSII$_a$ is greater (because of the presence of more antenna chlorophylls) than that of PSII$_\beta$. Little is understood of the mechanism of membrane stacking into grana and the biochemical and physiological significance of this phenomenon, although

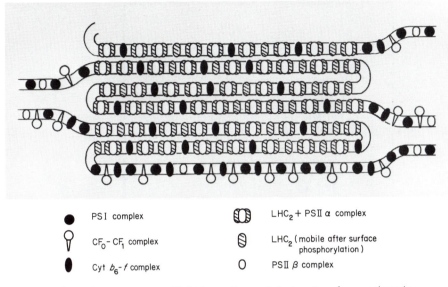

●	PS I complex	⬭	LHC₂ + PSⅡ α complex
ⵟ	CF₀- CF₁ complex	⬭	LHC₂ (mobile after surface phosphorylation)
⬮	Cyt b₆- f complex	○	PSⅡ β complex

Fig. 8.4   Schematic arrangement of light-harvesting and electron transfer complexes in the thylakoid membrane. (Courtesy: J. Barber, Imperial College, London.)

we know that the degree of stacking is influenced by the cation concentration and by the phosphorylation of light-harvesting proteins.

Thylakoid membranes can be mechanically disrupted by extrusion through a French or Yeda press and the smaller stromal vesicles in the fragments can then be separated from the larger granal stacks by differential centrifugation. The properties of the stroma lamellae thus isolated are different from those of isolated grana or whole thylakoids. For example, the 77 K fluorescence emission spectrum of the stroma lamellae shows a single peak at 735 nm due to PSI and LHC I, and almost no emission at 685 and 695 nm, characteristic of PSII (Fig. 8.5).

Recently, Albertsson's group have fractionated spinach chloroplasts by sonication and French press treatment and then separated the fractions by counter-current distribution or by aqueous–polymer phase partition (§4.7). The components of the various subfractions were identified by spectroscopic or immunological techniques. These studies show the existence of at least three domains in the thylakoid membrane: a circular domain in the centre of the grana containing mainly $PSII_a$, a peripheral annulus of the grana discs containing mainly PSI_a, and the stroma lamellae containing PSI_β and $PSII_\beta$. They also propose that the function of the grana is to support non-cyclic electron transport (ATP and $NADPH_2$ production and

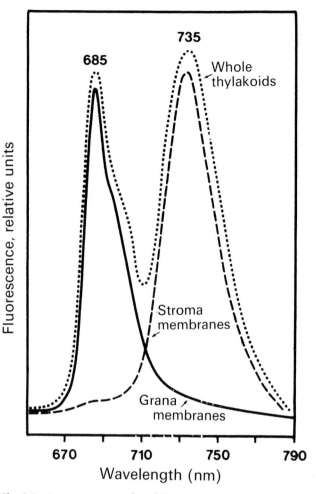

Fig. 8.5 Low-temperature (77 K) fluorescence emission spectra of isolated stroma and grana membranes compared with that of whole thylakoids. The 735 nm peak originates from LHCI attached to PSI, and the 685 and 695 nm peaks from LHC II attached to PSII. (Courtesy: D. Simpson.)

O_2 evolution), while the stroma lamellae are responsible for cyclic electron transport and phosphorylation.

Immunochemical studies show that the cyt f complex is located in three membrane domains: the appressed granal, the non-appressed stromal and end granal, and also in the non-appressed grana margins.

8.7   The photosystem II oxygen-evolving reaction

This key reaction of plant, algal and cyanobacterial photosynthesis is the main source of all O_2 on earth but still continues to be one of the great unsolved mysteries of photosynthesis. However, 'light' has begun to get through the 'black box', and recent work has significantly helped to probe the mechanisms involved in the oxidation of water. The problem is being attacked from two main directions: (1) isolation of PSII membranes with very good O_2 evolution rates and determination of the valence state and structure of the Mn atoms in the S_n states (§4.8) in the membranes by a variety of spectroscopic techniques; (2) synthesis of Mn clusters with spectroscopic characteristics that resemble the Mn spectra of PSII and which may even catalyze photo-oxidation of water. Classical spectroscopy such as absorption, infra-red (IR), nuclear magnetic resonance (NMR) and EPR are being supplemented by more advanced techniques of electron nuclear double resonance (ENDOR), electron spin echo, X-ray absorption spectroscopy, etc. to identify the paramagnetic Mn species generated in PSII during laser flash photolysis. Considerable progress has been made in identifying the polypeptides surrounding the Mn cluster of the water-oxidizing complex. The Mn ions are located in a compartment towards the lumenal side of the membrane.

The requirement of manganese in the photoxidation of water has been known since 1970. On the basis of atomic absorption measurements and EPR quantification of Mn ions released by acid treatment of PSII, it is now almost accepted that the active O_2-evolving complex contains four atoms of Mn per P680 (although a few researchers believe there are six Mn per P680). Treatment with 0.8 M Tris buffer pH 8.5 releases most of the Mn from the complex, with the concomitant loss of the three polypeptides EP33, EP24 and EP17 of the regulatory cap. The Mn complex can be reconstituted in light; the kinetics of reassembly of the Mn tetramer appear to be a multistep process.

We have already mentioned (§4.8) that in 1970 Joliot, and Kok, proposed the S state model of charge accumulation on the water oxidation catalyst to explain why the yield of O_2 from chloroplasts in response to a sequence of saturated light flashes oscillates with a period of 4. Most of the studies in recent years have tried to correlate the Mn signals from spectroscopic data with the S states of the water-oxidizing complex. Two EPR signals observable at very low temperatures (< 35 K) have been attributed to the S_2

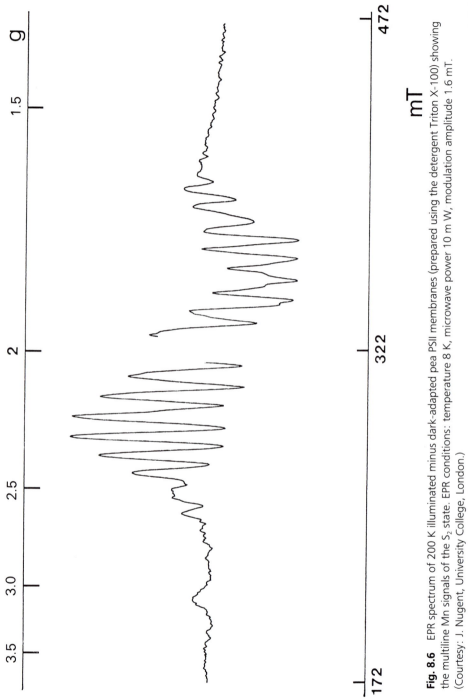

Fig. 8.6 EPR spectrum of 200 K illuminated minus dark-adapted pea PSII membranes (prepared using the detergent Triton X-100) showing the multiline Mn signals of the S_2 state. EPR conditions: temperature 8 K, microwave power 10 m W, modulation amplitude 1.6 mT. (Courtesy: J. Nugent, University College, London.)

state. These are known as the multiline EPR signals with $g = 2.0$ and the $g = 4.1$ signal (Fig. 8.6). The signals are perturbed by treating the samples with inhibitors of O_2 evolution, e.g. ammonia ions or NH_2OH. These EPR signals suggest that the Mn ions have a mixed valence in the S_2 state; however, definite assignment of the Mn oxidation state is difficult. The four Mn atoms may be arranged as a tetramer (or a pair of interacting dimers) or as a trimer which gives rise to the two EPR signals and an EPR-silent monomer. In dark-adapted PSII membranes, an EPR signal with $g = 4.8$ has been detected and has been proposed to arise from a precursor of the multiline signal and assigned to the S_1 state.

Valuable information regarding the ligands of Mn ions and Mn–Mn, Mn–O and Mn–N distances in the cluster has been obtained from X-ray absorption studies. These spectroscopic probes suggest that the atomic environment of Mn ions is heterogeneous. Flash-induced absorption changes in the UV that oscillate with a period of 4 have been assigned to the individual S state transitions. These absorption changes show correlations with the changes in the valence state of Mn from MnII to MnIII to MnIV. Spectroscopic data suggest that the Mn ions undergo oxidation during the $S_0 \longrightarrow S_1$ and $S_1 \longrightarrow S_2$ transitions, but it is still not clear whether Mn is oxidized on the $S_2 \to S_3$ transition.

Location of the Mn cluster   Studies with wild-type and mutant algae show that the EP33 polypeptide (Fig. 8.7) may be required for the assembly and activation of the Mn cluster and to optimize catalytic activity. Chemical modification of various amino acid residues in PSII polypeptides and subsequent measurement of the O_2 evolution rate and spectral properties of the modified PSII suggest that residues in the D_1–D_2 polypeptide (possibly histidine and carboxylate) may provide ligands for the Mn cluster. Similar conclusions were arrived at from site-directed mutations of amino acid residues in D_1 and D_2 polypeptides in the cyanobacterium *Synechocystis*.

The mechanism of water oxidation is still unknown. It is assumed that the Mn ions form the matrix to which two water molecules co-ordinate to form a dioxygen bridge. There are two possibilities for the chemistry of the oxidation. (1) A concerted four-electron oxidation occurs at the $S_3 \longrightarrow S_4 \longrightarrow S_0$ transitions. It is generally believed, but not unequivocally established, that water oxidation occurs at the $S_4 \longrightarrow S_0$ transition. (2) A two-step reaction with the formation of a peroxide intermediate prior to the $S_3 \longrightarrow S_4 \longrightarrow S_0$ transition. Sequential one-electron oxidation of water with the formation of OH^- intermediates is considered to be energetically unfavourable.

The roles of Ca^{2+} and Cl^- in the water oxidation are not definitely established. Atomic absorption measurements have shown that PSII membranes contain two to three bound Ca ions per P680. Bound Ca can be released with a loss of O_2 evolution by washing the membranes with 1–2 M NaCl supplemented with 1–5 mM EDTA, in light. Measurement of O_2 evolution of Ca-depleted PSII membranes as a function of added Ca has shown that PSII contains one high-affinity and two low-affinity Ca-binding sites per P680. X-ray absorption spectroscopy has indicated that Ca ions are located very close to the Mn complex. Whether the role of Ca is only to protect the Mn complex or whether it directly participates in the O_2 evolution reaction is still a moot question. The influence of Ca ions in electron transfer and O_2 evolution in cyanobacteria may be different from that in plants.

The binding of Cl ions to PSII requires the prior binding of Ca. Binding values were attained by NMR spectroscopy using ^{35}Cl and isotopic labelling with ^{36}Cl. Reported values for the number of Cl^- per P680 required for maximum O_2 evolution have varied from 1 to 5. Some EPR measurements indicate that Cl^- is essential for the $S_2 \longrightarrow S_3$ transition. Requirement for Cl^- is not specific; it can be replaced by Br^-.

From the foregoing discussion it is obvious that we still need to learn a lot more about the location, structure, chemical reactions, and the role of cofactors of the water oxidation complex in order to elucidate the mechanism of O_2 evolution.

8.8   Photosystem II: structure and function

Photosystem II is that part of the oxygenic photosynthetic apparatus which catalyzes the light-induced oxidation of water coupled to the reduction of plastoquinone. This function can be represented by the simple equation:

$$2H_2O + 2PQ \xrightarrow[\text{PSII membranes}]{\text{light}} O_2 + 2PQH_2$$

During this process, four photons are transferred across the thylakoid membrane from the stromal side to the lumenal side which are utilized in ATP synthesis. PSII is located within the thylakoid membrane in close association with lipids.

Major advances have been made over the last decade towards our understanding of the structure and function of PSII. These include: (1) the

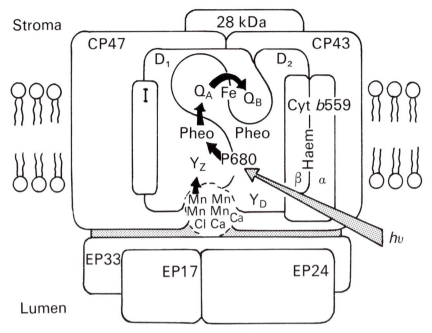

Fig. 8.7   Simplified scheme showing the main components of the PSII complex and the electron transfer sequence in the complex. For details refer to the text. (After Debus, 1992.)

isolation of almost pure PSII particles, free from other thylakoid components; (2) separation and identification of the polypeptides in the purified complex by gel electrophoresis; determination of the primary structure of these polypeptides by isolating the genes, then cloning and sequencing them; (3) most importantly, the determination of the structure of the reaction centres of photosynthetic purple sulphur bacteria and the observation that the PSII and purple bacterial reaction centres are structurally and functionally related.

Based on the information available at present, a structural model for PSII can be proposed as in Fig. 8.7. There are three functional parts: (1) an antenna consisting of pigment–protein complexes, (2) a reaction centre containing the primary reactants, plastoquinones, and the water-oxidizing complex, and (3) a regulatory cap made up of polypeptides extrinsically bound to the lumenal surface of the thylakoid membrane.

The *light-harvesting antenna* of PSII can be further subdivided into proximal and distal antennae. The proximal antenna, which surrounds the P680 core and is tightly coupled to it, consists of two pigment–protein complexes, CP47 and CP43. (CP = chlorophyll–protein complex.) Both complexes contain Chl a and β carotene bound to the apoproteins, but no Chl b. Excitation energy captured by the bulk antenna (proximal as well as distal) is transferred to P680 from the proximal antenna. The distal antenna is mainly composed of light-harvesting complexes (LHC II) and contains covalently bound Chl a, Chl b and xanthophylls. The three-dimensional structure of LHC II at 6 A^0 resolution has recently been determined by electron crystallography. This low-resolution structure shows that LHC II forms a trimer, with each monomer containing 15 chlorophylls and several carotenoids. In addition to its function in light harvesting and energy transfer to P680, the LHC II also plays a role in energy distribution between the two photosystems by undergoing reversible phosphorylation–dephosphorylation (§8.14).

The *regulatory cap* is comprised of a set of hydrophilic extrinsic proteins (EP) that have apparent molecular masses of 33, 24 and 17 kDa and are designated as EP33, EP24 and EP17 respectively. The EP33 acts as a barrier between the water-oxidizing unit attached to the membrane and the aqueous lumenal phase. Spectroscopic and protein cross-linking studies suggest that EP33 is in contact with the reaction centre. It may also be required for the assembly and activation of the Mn complex in eukaryotic algae and plants *in vivo*.

The *PSII reaction centre* can be functionally subdivided into the light-absorption and electron transfer segments and the water-oxidizing complex. The PSII *core* is the minimal unit able to perform light-induced charge separation and electron transport (but not to evolve O_2), and in this respect resembles the purple bacterial reaction centre. This core has been isolated now in many laboratories and comprises 4 to 6 Chl a, 2 Pheo, 2 β carotene, 1 D_1–D_2 heterodimer, 1 CP47, 1 CP43, 1 (or 2) cytochrome b559 and 1 10 kDa polypeptide (I) per P680. No quinone (Q_A) is present in the PSII core unit. The P680 and Pheo are believed to be bound to the D_2–D_1 (so called because they were first identified as two *diffuse* bands by gel electrophoresis and staining) proteins. In O_2-evolving PSII reaction centre complexes, the quinones (Q_A and Q_B) and ferrous iron are also bound to the D_2–D_1 heterodimer.

The primary charge transfer from excited P680 to Pheo takes place in 3 picoseconds. Recombination of the charged species is prevented by the rapid (250–500 μs) oxidation of Pheo$^+$ by a neighbouring Q_A and the instant reduction of P680$^-$ by a redox active tyrosine residue Yz of the D_1 polypeptide.

$$\text{Pheo}^- + Q_A \longrightarrow \text{Pheo} + Q_A^-$$

Q_A^- is oxidized within 100 to 200 μs by a second plastoquinone, Q_B, forming a Q_B semiquinone. This semiquinone accepts another electron from a different Q_A^- to form Q_B^{--}. (Note that Q_A is a single-electron acceptor while Q_B is a two-electron acceptor).

$$Q_A^- + Q_B \longrightarrow Q_A + Q_B^-$$
$$Q_B^- + Q_A^- \longrightarrow Q_A + Q_B^{--}$$

By analogy with the known sequences of the redox events occurring in purple bacterial reaction centres, it is believed that the fully reduced Q_B^{--} accepts two protons from the stroma, forms $Q_B H_2$, and leaves the reaction centre as plastoquinol and then transfers the protons to the cyt bf complex (Fig. 5.2). The Q_B site in the D_1 protein is then occupied by another Q_B molecule from the plastoquinone pool.

There is another redox active tyrosine, Y_D in the D_2 polypeptide. Its function, if any, is unknown. Both Y_Z and Y_D were detected by EPR spectroscopy and their precise locations in the polypeptides were identified by site-specific mutagenesis.

Both O_2-evolving PSII preparations as well as the PSII core contain cytochrome b559, a heterodimer of 9-kDa (α) and 4-kDa (β) polypeptides co-ordinated to a haem. The possible functions which have been assigned to cyt b559 are as a photoprotective agent and as a proton pump during cyclic electron flow around PSII.

8.9   Photosystem I

Photosystem I, in higher plants, is a membrane-bound protein complex which binds three different types of components, viz. the reaction centre (P700), the photoreducible electron acceptors, and the light-harvesting antenna (Fig. 8.8). There are approximately 200 chlorophylls per reaction centre (see Fig. 3.12). Functionally, PSI catalyzes the light-driven transfer of electrons from reduced plastocyanin, which is located on the lumenal side of the thylakoid membrane, to ferredoxin, the soluble protein which is located on the stromal side of the membrane. The ferredoxin, in turn, reduces NADP to $NADPH_2$ in the presence of the enzyme ferredoxin–NADP reductase (FNR), which is also located on the stromal side of PSI. Plastocyanin and ferredoxin are held close to the PSI membrane complex by electrostatic attraction towards specific protein subunits of the complex. These subunits are usually referred to as the plastocyanin docking (PSI–F) and the ferredoxin docking (PSI–D/PSI–E) subunits (Fig. 8.8) respectively.

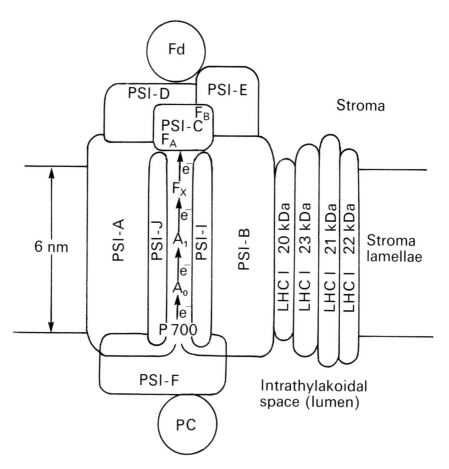

Fig. 8.8 Components of, and electron transfer in, PSI. (For details see text and Tables 3.2 and 8.1.)

The technique of electron paramagnetic resonance (EPR) spectroscopy has been used to full advantage in identifying many of the electron mediators (Fig. 8.9) and the pathway of electron transfer in the PSI complex. The findings derived from EPR and other spectroscopic investigations have been confirmed by SDS-PAGE of PSI membrane proteins, by protein cross-linking studies, and by gene cloning and sequencing. Some of the subunits of the PSI complex are given in Table 8.1.

The reaction centre chlorophyll P700 is a dimer. The primary acceptor A_0 is a special Chl a; its optical absorption spectrum (oxidized minus reduced) is slightly different from that of normal Chl a. The exciton transfer

Table 8.1. *Some subunit proteins of PSI complex (for details refer to Ikeuchi, 1992)*

Component	Molecular mass (kDa)	Gene	Location	Function
PSI-A	81–84	psa A(C)	Transmembrane	Binding site for Chl a, P700, A_0, A_1 and F_x
PSI-B	81–83	psa B(C)	"	"
PSI-C	9	psa C(C)	Peripheral	Binding site for F_A and F_B
PSI-D	15–18	psa D	"	Ferredoxin-docking
PSI-E	8–11	psa E	"	Ferredoxin-docking
PSI-F	16–19	psa F	Transmembrane & peripheral	Plastocyanin-docking
PSI-I	4–5	psa I(C)	Transmembrane	?
PSI-J	4–5	psa J(C)	Transmembrane	?

Notes: The protein subunits are usually separated by SDS–PAGE of the PSI complex and identified by protein sequencing and/or gene cloning techniques. Genes denoted by (C) are coded by the chloroplast, and others by the nuclear genome.

from P700 to A_0 and the charge separation occur at a very fast rate (15 picoseconds) and with almost 100% efficiency. The secondary acceptor A_1 is most probably a phylloquinone (vitamin K_1), as deduced from its optical and EPR spectra. However, the role of phylloquinone in the electron flow from A_0 to the Fe–S centres is not unequivocally proven. F_x, based on its EPR properties, is a [4Fe–4S] centre bound to protein with an extremely negative redox potential ($E_m = -730$ mV). A_0, A_1 and F_x are all bound to the PSI–A/PSI–B heterodimer. F_A and F_B are also [4Fe–4S] centres bound to protein subunits. There are four protein subunits binding Chl a/b in the LHCI antenna; these subunits can be removed from the PSI complex by treatment with non-ionic detergents.

The path of electron transfer in PSI of cyanobacteria is similar to that of higher-plant PSI; the electron donor to P700 in some cyanobacterial species is a cytochrome, cyt-553, instead of plastocyanin. The chlorophyll content of PSI in cyanobacteria is less (approximately 100 Chl per RC) than that of PSI of higher plants.

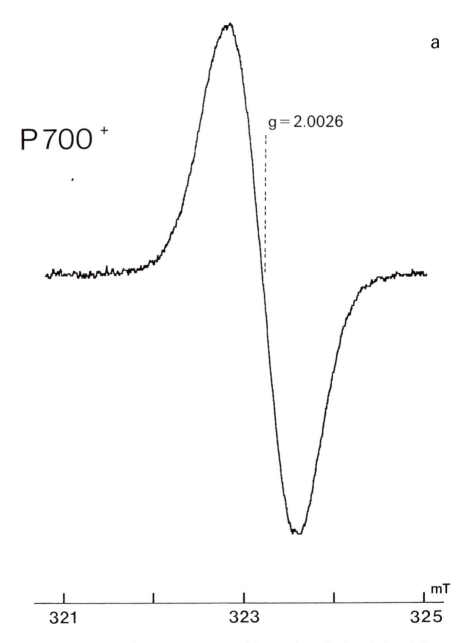

Fig. 8.9   EPR spectra of some PSI components. (**a**) P700[+] (15 K illuminated minus dark of PSI particles). (**b**) Photo reduced Fe–S centre A in 15 K illuminated minus dark PSI particles (**top**), photoreduced Fe–S centres A and B conjugated to isolated 9-kDa

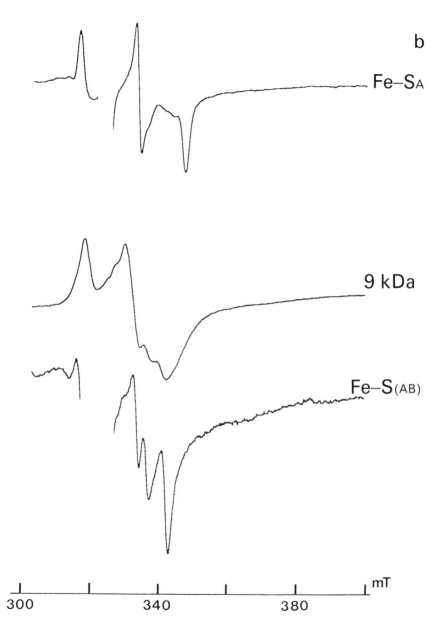

b

Fe–S_A

9 kDa

Fe–S(AB)

mT

300   340   380

polypeptide (psa *a* gene product), and dithionite reduced Fe–S centres A and B which
show their interaction (**bottom**). (Courtesy: J. Nugent.)

8.10   The cytochrome b_6f complex: the Q cycle

The cyt b_6f complex transfers electrons from PSII to PSI during linear electron flow from water to NADP (see Fig. 5.2). It can be considered as a plastoquinol–plastocyanin oxidoreductase. Recent work by J. Anderson (in Canberra) and others suggests that the cyt b_6f complexes are distributed in three domains of the thylakoid membrane: the appressed granal stacks (along with PSII), the non-appressed stromal thylakoids (along with PSI), and the non-appressed grana margins.

The b_6f complex can be isolated from chloroplasts by detergent fractionation and then further resolved into individual protein components by treatment with denaturing agents. The isolated complex contains four major polypeptides: cyt b_6 (molecular mass 23 kDa, *pet* B gene product), cyt f (31 kDa, *pet* A), the Rieske (discoverer) iron–sulphur protein FeS_R (20 kDa, *pet* C), and subunit IV (17 kDa, function unknown). The electron transfer sequence within the complex is: $PQH_2 \rightarrow FeS_R \rightarrow cyt\ f \rightarrow PC \rightarrow PSI$. Cyt f is held in the thylakoid by a single transmembrane span located near its hydrophobic C-terminus, with a globular hydrophilic segment protruding into the lumen. Concomitant with electron transfer, the complex also translocates protons from the stromal to the lumenal side of the membrane. Other functions of the b_6f complex are electron transfer from reduced Fd to PC via b_6 during cyclic photophosphorylation, and possibly in the regulation of light energy distribution between the two photosystems (in response to ATP demand) via LHC II phosphorylation–dephosphorylation reactions. It has been postulated that the lateral movement of phosphorylated LHC II from the grana ($PSII_a$) membranes is accompanied by a simultaneous movement of the cyt b_6f from the grana to the stromal thylakoids ($PSII_\beta$ and PSI).

The mechanism of proton translocation across the thylakoid membrane mediated by the b_6f complex is still not understood. There are many similarities (and differences) in the structure, composition and function of the chloroplast b_6f complex and the bc complex in the mitochondrial respiratory chain of mammals and bacteria. Cytochrome f, so named because it was detected by its unique absorption at 555 nm in leaves (Latin – *frons*), is truly a c-type cytochrome. Mitchell had proposed a *Q cycle* to explain the mechanism of proton translocation via the mitochondrial ubiquinone–bc_1 complex, the key step being an electrogenic electron transfer through the site of quinol oxidation. Does a similar Q cycle operate in the chloroplast cyt b_6f complex? The electron and protein transfer across the membrane can be measured from light-induced changes in the absorption at 320 nm due to proton uptake. The participation of the b-type haems in the reaction sequences can be detected from the electrochromic absorption changes at 515/524 nm that occur when thylakoids are illuminated by single turnover flashes.

Proponents of the Q cycle operation propose that during electron transport FeS_R accepts one electron from PQH_2 (Q_B doubly reduced by PSII picks up two stromal protons and forms PQH_2) and donates it to cyt f. Simultaneously, PQH_2 transfers its second electron to one of the b haems (low potential) and releases the two protons to the lumen. The electron from the low potential b haem is transferred to PQ via the high potential b haem of the cyt b_6 (Fig. 8.10). The PQ gets doubly reduced after two cycles of electron transfer; doubly reduced PQ picks up two more stromal protons and enters the quinone pool.

Although some authors believe that the b_6f complex functions electrogenically during electron transport, there are others who consider the evidence as equivocal. Furthermore, if the Q cycle is truly operative, $2H^+$ should be translocated per electron transferred to cyt f; experimentally determined values of H^+/e^- range from 2 to 3. The mechanism of proton translocation across the thylakoid membrane is still an open question; maybe a modified catalytic Q cycle or a proton pump is involved in the translocation.

8.11   RuBisCO: structure and function

Ribulose bisphosphate carboxylase-oxygenase (RuBisCO) is probably the most abundant protein on earth; it constitutes about half of the leaf proteins. RuBisCO catalyzes the biological fixation of 10^{11} tonnes of CO_2 annually from the atmosphere. In the last two decades considerable research effort has been expended towards the elucidation of the mechanism of carboxylation by the enzyme and in the determination of its three-dimensional structure. Besides fundamental understanding, a main objective behind this research is to design, if at all possible, a RuBisCO molecule with an increased carboxylation to oxygenation activity ratio than is found in the naturally occurring enzyme.

RuBisCO isolated from oxygenic photosynthetic organisms is an octamer consisting of eight large subunits (LSU, molecular mass 50–55 kDa) and eight small subunits (SSU, molecular mass-12–15 kDa). The LSU, which is coded by the chloroplast DNA, is synthesized within the

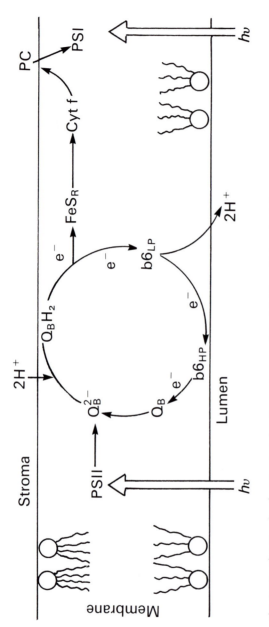

Fig. 8.10 The Q cycle of proton translocation; a proposed model. $b6_{HP}$ = high potential b haem; $b6_{LP}$ = low potential b haem. See text for details.

chloroplasts; the catalytically active site(s) is located in the LSU. The SSU is coded by the nuclear genome and is assembled in the cytoplasm as a preprotein that contains a domain of 50 extra amino acids at the aminoter-minal end; this domain is cleaved off when the preprotein enters the chloroplast through the envelope. Assembly of the matured SSU and LSU within the chloroplast is mediated by a *chaperone*, the RuBisCO assembling protein. (For subunit arrangement see Plate IV.).

In contrast to the RuBisCO enzymes of chloroplast stroma, the function-ally active RuBisCO of *Rhodospirillum rubrum* is a dimer of 50 kDa subunits which lack any SSU. The X-ray structure of *R. rubrum* enzyme has also been determined, and many residues in the enzyme molecule have been altered by site-directed mutagenesis. These studies have helped to identify the activator and substrate-binding sites on the RuBisCO molecule.

Activation of RuBisCO   Prior to catalysis (either carboxylation or oxyge-nation), RuBisCO has to be converted from an inactive state to an active carbamylated form by condensation with a molecule of CO_2 and by subsequent addition of a divalent metal ion:

$$E + CO_2 \xrightarrow{-H_2O} E.CO_2 \xrightarrow{+Mg^{2+}} E.CO_2.Mg^{2+}$$

inactive enzyme   Carbamate   active enzyme

The carbamylation (condensation between CO_2 and enzyme $-NH_2$ group) occurs at a specific Lys residue at the activator site of the enzyme which is distinct from the catalytic site. Activation of RuBisCO also requires light. Recent work with a mutant of *Arabidopsis* (a plant amenable to mutations and genetic engineering) has led to the discovery of a protein named *rubisco activase* which is involved in the light-regulated activation of RuBisCO.

The action of rubisco activase is dependent on the availability of ATP and a membrane pH gradient, which are generated during light-induced electron transport. Many plant species produce 2-carboxy arabinitol-phosphate, an analogue of a six-carbon intermediate formed during RuBisCO catalysis, which inhibits the enzyme by occupying the catalytic site. It has been proposed that rubisco activase functions by removing this inhibitor from the catalytic domain of RuBisCO during transition from darkness to light. RuBisCO is one of the light-regulated chloroplast enzymes not controlled by the Fd–thioredoxin system.

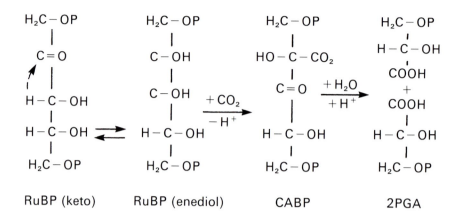

Fig. 8.11 Carboxylation of ribulose bisphosphate catalyzed by RuBisCO. CABP, carboxyarabinitol phosphate.

Catalysis by RuBisCO   The catalysis takes place in three steps. (1) Translocation of a proton from the C_3 position of the substrate RuBP to the C_2 position and formation of an enediol. (2) The enediol reacts with a molecule of CO_2, forming the intermediate adduct 3-keto-2-carboxy arabinitol bisphosphate (CABP). This intermediate can be isolated in a stable form. Crystal structure analysis of the quaternary complex of spinach RuBisCO co-ordinated to CABP has confirmed the nature of the substrate-binding sites on the enzyme (Plate IV). These had been previously deduced from comparative sequence analysis, chemical modifications, and site-directed mutagenesis. The residues involved are thought to be a specific Lys and Glu. (3) The intermemdiate is hydrolytically cleaved into one molecule of PGA and a C_2 carbanion; the carbanion is immediately protonated to yield the second molecule of PGA (Fig. 8.11).

Athough many derivatives of naturally occurring RuBisCO have been prepared by chemical or genetic modification of the molecule, none of these derivatives has manifested an enhanced ratio of carboxylation to oxygenation activity compared to the native enzyme.

As mentioned in §8.3, RuBisCO is one of the proteins subjected to extensive gene sequence determination in recent years. The genetic dissection of the enzyme in the green alga *Chlamydomonas reinhardtii* had identified three large-subunit (L) regions that should provide loci for future investigations via genetic engineering.

8.12   Fluorescence as a probe for energy transfer and stress physiology in photosynthesis

It was mentioned in Chapter 4 that part of the energy quanta absorbed by chlorophyll is emitted as fluorescence in the red. In green photosynthetic tissues this fraction of excitation energy which is dissipated as fluorescence *in vivo* is small (3–5%); in solution, however, depending on the solvent, this fraction may reach up to 30%. Fluorescence in photosystems is a measure of absorbed quanta that are not utilized in photosynthesis.

Fluorescence spectral characteristics, such as the nature and intensity of the emission bands, lifetime, quantum yield, and induction kinetics, all reflect the properties of chlorophyll molecules and their environment and can thus be used to study photosynthetic electron transport and associated physiological processes. Here we discuss only the basics concerned with fluorescence induction curves and their application in oxygenic photosynthesis.

The fluorescence induction curve (the Kautsky effect)

The time course of chlorophyll *a* fluorescence emission in photosynthetic materials is characterized by the fluorescence induction curve (Fig. 8.12). If a leaf or chloroplast preparation kept in darkness for some period (say 30 minutes) is illuminated with a flash of saturating light, its fluorescence rises steeply to an initial level 0 (fluorescence F_0) and then rises within fractions of a second to a peak P (maximum fluorescence, F_m). It then drops in seconds to a steady state value S (F_s). This sudden rise and subsequent slow drop of chlorophyll *a* fluorescence, and its relation to photosynthetic activity, was first observed by Kautsky in 1931 in Germany and is known as the *Kautsky effect*. The nature of the fluorescence induction curve is dependent upon the chlorophyll content of the material, the period of prior adaptation, the intensity of actinic illumination, the presence of inhibitors, etc., and it can be qualitatively and quantitatively slightly different from the simple curve in Fig. 8.12 illustrating the phenomenon.

Now, let us interpret the fluorescence induction curve in terms of the photosynthetic processes, particularly to explain the events occurring in photosystem II. During dark adaptation in leaves there is no chance for photochemistry to occur. The initial steep rise in fluorescence, F_0, which happens in pico to nanoseconds, mainly originates from the antenna

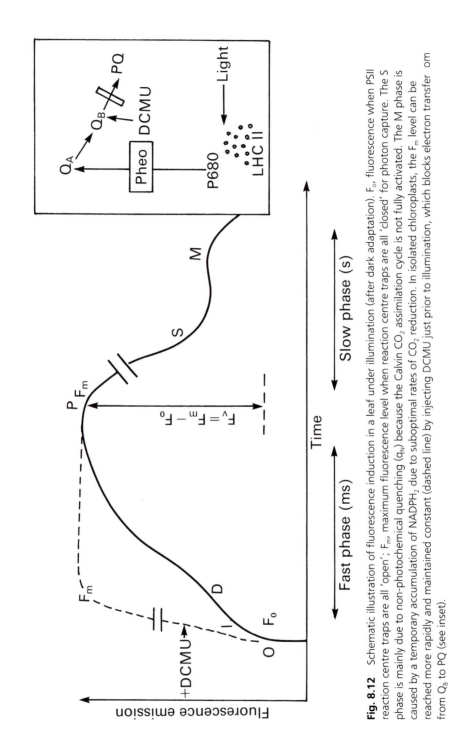

Fig. 8.12 Schematic illustration of fluorescence induction in a leaf under illumination (after dark adaptation). F_o, fluorescence when PSII reaction centre traps are all 'open'; F_m, maximum fluorescence level when reaction centre traps are all 'closed' for photon capture. The S phase is mainly due to non-photochemical quenching (q_N) because the Calvin CO_2 assimilation cycle is not fully activated. The M phase is caused by a temporary accumulation of $NADPH_2$ due to suboptimal rates of CO_2 reduction. In isolated chloroplasts, the F_m level can be reached more rapidly and maintained constant (dashed line) by injecting DCMU just prior to illumination, which blocks electron transfer om from Q_B to PQ (see inset).

chlorophylls of PSII (with only a minor contribution from PSI antenna) before the excitons are transferred to the reaction centre. At level 0, all the reaction centre chlorophylls are 'open' to accept photons and all the electron mediators in PSII are in the oxidized state, i.e. F_0 measures the fluorescence of open reaction centres. Once the reaction centre chlorophylls are excited, charge transfer begins, Q_A becomes reduced and the fluorescence rapidly rises to an inflection point I (Fig. 8.12). Then, as electrons are transferred from reduced Q_A to Q_B, the rate of increase in fluorescence decreases (point D). When all the Q_A molecules are reduced, there can be no further transfer of charge from the reaction centre chlorophylls to Q_A and so the centres cannot accept any more excitations, i.e. they are closed. The probability of dissipation of excitation energy from chlorophyll a is highest at this stage and it is reflected by the peak P in the induction curve with the fluorescence maximum, F_m. The subsequent 'slow' decay of fluorescence from P to S results from the oxidation of Q_A as electrons are transferred away to Q_B and then beyond to NADP via the cyt bf complex and PSI, and also partly due to non-photochemical quenching (see later).

Application of fluorescence induction data in photosynthesis   The variable fluorescence rise from 0 to P (see Fig. 8.12) that originates from Chl a of phosystem II photochemistry is related to the rate of electron transfer from the reaction centre to Q_A. The variable fluorescence F_v is the difference in fluorescence level between its maximum, F_m, when all the Q_A molecules are fully reduced (closed reaction centres), and its minimum level F_0, when all the Q_A are in the oxidized state (open reaction centres). This difference, F_v, $= (F_m - F_0)$, is also referred to as *photochemical fluorescence quenching*.

It has been shown that the photochemical yield of photosystem II is equal to the ratio between F_v and F_m (i.e. $F_v{:}F_m$). This ratio can be measured accurately at liquid N_2 temperature (77 K) since primary photochemical reactions are independent of temperature while other physiological events are stopped at 77 K. For healthy leaves from a large number of vascular plants, Bjorkman and Demmig-Adams (in California) have found the $F_v{:}F_m$ ratio to be around 0.83. It has been verified experimentally that $F_v{:}F_m$ is proportional to the rate of photosynthesis measured as O_2 evolution or CO_2 assimilation. The value of $F_v{:}F_m$ is affected by stress factors, particularly by photoinhibition.

As a result of the advances made in the last 5 years in the techniques for measuring room and field temperature fluorescence of leaves and algae, and in the interpretation of the fluorescence data, it is now possible to evaluate the quantum yield of electron flow through PSII *in vivo*. Equipment to measure PSII fluorescence arising from a very low-intensity modulated light, while the leaf is performing photosynthesis in higher actinic light (modulated fluorescence measurement), is now available (Fig. 1.7 and Plate VII). To evaluate the quantum yield of PSII electron flow (ϕ_{II}), two measurements of modulated fluorescence are made to determine (1), $(F_m)_s$, the steady state fluorescence measured over a long period (minutes) under the given environmental conditions (light, temperature, CO_2 level etc.), and (2) F_m, the maximum fluorescence emitted during a saturating flash of light (<1 s duration) under the same conditions. Based on the assumption that ϕ_{II} is proportional to the product of the fraction of open centres and the efficiency of excitation capture by the open centres, the value of ϕ_{II} has been calculated by Genty *et al.* (*Biochim. Biophys. Acta*, 1989) as $\phi_{II} = \{F'_m - (F_m)_s\}/F'_m$. The theoretical maximum for ϕ_{II}, when all the reaction centre traps are open, should be 1. However, experimentally the values have not exceeded 0.85.

Photochemical and non-photochemical fluorescence quenching   Photochemical fluorescence quenching arises from charge separation at photosystem II reaction centres. The coefficient of fluorescence quenching by photochemistry, q_P, indicates that proportion of photons absorbed by photosystem II that is used up by the open reaction centres for photosynthesis. It is almost the same as $F_v:F_m$. The slow rise in the fluorescence kinetics curve is correlated with non-photochemical fluorescence quenching q_N. In intact leaves it was difficult to distinguish between the contributions of q_P and q_N to the total fluorescence of the slow phase. However, using the technique of modulated fluorescence, it is now possible to separate q_P and q_N in illuminated leaves. Factors contributing to q_N are the generation of a strong pH gradient in the thylakoid membrane, phosphorylation and dephosphorylation of light-harvesting protein complex and associated energy redistribution, the concentration of cations (Mg^{2+}), photoinhibition, the development of a zeaxanthin cycle, etc. All these factors contribute to the energy-dependent quenching (q_E) which is a major fraction of q_N.

Fluorescence data by themselves cannot be fully relied on as a measure of phosynthetic efficiency of a plant; they can, however, be used in conjunction with other parameters such as oxygen evolution measured polarographically, and CO_2 assimilation measured by IRGA. The technique of fluorescence measurement and the interpretation of fluorescence data have advanced rapidly in the past few years, although there is not complete agreement in the interpretation of photosynthetic fluorescence phenomena.

Many researchers now use fluorescence emission as a yardstick for measuring photosynthetic productivity of individual plants and canopies.

8.13   Photoinhibition

The photosynthetic efficiency of many plants is decreased when they are subjected to stress conditions such as high light intensity, low temperature, low CO_2 environment, water limitation, etc. Photoinhibition is defined as the decrease in photosynthetic capacity induced by the exposure of photosynthetic tissues to high fluxes of photosynthetically active radiation (400–700 nm). Not included in this definition of photoinhibition are inhibition of photosynthesis caused by exposure to UV radiation, and photo-oxidation (which usually follows photoinhibition) leading to the bleaching of photosynthetic pigments. In most cases, photoinhibition is reversible – normal photosynthetic rates are restored, after a lag period, when the organism is exposed to less intense irradiance. Ultrastructure studies using the electron microscope reveal no permanent damage to the thylakoid membranes during photoinhibition except where prolonged stress occurs. The chloroplast is usually able to regenerate any damaged components. Photoinhibition in plants has been observed under three types of conditions: when a plant is exposed to higher irradiance than that under which it has been grown, when the plant is subjected to conditions which decrease its rate of carbon metabolism, and when some species are exposed to chilling temperatures (10°C and below), even under normal irradiance.

The extent of photoinhibition can readily be determined by measuring photosynthetic electron transport as O_2 exchange or fluorescence (induction and decay) which reflect electron transfer around the two photosystems. Photoinhibition results in a decrease of quantum yield which is related to the PPFD to which the plants are exposed. In isolated chloroplasts, photoinhibition is much more evident in PSII activities than in PSI. Measurement of photosynthetic electron transport through various segments of chloroplast membranes isolated from photoinhibited algae and plants has shown a definite correlation between photoinhibition and the inactivation of the D_1 polypeptide which binds Q_B, a PSII electron mediator. Photoinhibition in PSI may be due to an accumulation of electrons over and above those which can be transferred to NADP.

Comparative analysis of some leaf enzymes from chilling-sensitive and chilling-resistant plants has indicated a lower content of superoxide dismutase and catalase, two enzymes involved in oxygen metabolism, in the chilling-sensitive strains.

Mechanism of photoinhibition

Two different hypotheses have been put forward to explain the mechanism of photoinhibition: (1) damage to the reducing side of PSII at the Q_B or Q_A binding site; (2) damage to the oxidizing (water-splitting) side of PSII which prevents a stable charge separation; the site of the photoinhibitory effect could be P680 or Y_Z, the electron donor to P680. Thylakoid membranes exposed to high light intensity release fragments of D_1 protein which can be identified by gel electrophoresis. This degradation of D_1 protein is due to a PSII membrane-bound protease, whose activity is probably triggered by light. Concomitant with the cleavage of D_1 protein, there is a release of free Mn ions and of the extrinsic polypeptides (EP33, EP24 and EP17; see Fig. 8.7) into the lumen; all these factors contribute to the lowering of the rate of O_2 evolution.

If a leaf under illumination is given a pulse of ^{35}S Met (the radioactive form of the amino acid methionine) and the PSII fraction isolated from that leaf is subjected to SDS-PAGE and autoradiography, the main band that becomes visible radioactively is that of the D_1 protein. Thus, the D_1 protein has the fastest turnover among the chloroplast proteins. This *de novo* synthesis of the D_1 polypeptide occurs in the non-appressed regions of the thylakoid to which the ribosomes have access, while most of the functional PSII is located in the appressed regions. The newly assembled D_1 polypeptide then migrates to the appressed region. A lateral migration of PSII units from the appressed to non-appressed regions of the thylakoids has been observed during the release of D_1 proteins. Under normal illumination, the rate of *de novo* synthesis of the D_1 protein more than balances its degradation, and the leaf does not show any visible effects of photoinhibition.

Under high light exposure of PSII there is a rapid flux of electrons from $Pheo^+$ to Q_A (bound to the D_2 protein), and it is possible for the Q_A to be reduced beyond its normal Q_AH (one electron) level to Q_AH_2 (two electrons). This would temporarily block electron transfer from the RC to Q_A, and a P680–Chl triplet state will be generated at the reaction centre which, under aerobic conditions, would react with molecular O_2 forming singlet oxygen, $O^.$ (Fig. 6.13). This highly active oxygen species can bleach P680 and also trigger the degradation of the D_1 polypeptide.

Another condition producing photoinhibitory effects occurs when the rate of transfer of excitons to the PSII RC exceeds the rate of electron withdrawal from the RC by the primary electron acceptor Q_A. This results in a build up of $P680^+$ which has sufficient positive potential to oxidize and destroy Chl 670 (of the LHC) and β carotene.

Repair mechanisms

1. *Protection by the reaction centre core of PSII*   There are two β carotenes within the D_1–D_2 heterodimer which might quench chlorophyll triplets and deactivate toxic oxygen species. During photoinhibitory treatment, cyt $b559$ undergoes redox changes and is converted from a high potential to a low potential form. It is postulated that cyt $b559$ may mediate electron transfer from reduced plastoquinones to P680 in a cyclic flow and offer some photoprotection. Since the F_v:F_m fluorescence ratio (§8.12) is a fairly accurate measure of the PSII photosynthetic efficiency, the decrease of this ratio when plant materials are exposed to high light gives an estimate of the degree of photoinhibition of PSII.

2. *Repair mechanisms associated with the capture and delivery of light energy to PSII reaction centre*   Photoinhibition is caused not by high light *per se*, but rather by excess of light energy absorbed above the photosynthetic capacity of the organism. Many plants have evolved molecular strategies to acclimatize and protect themselves against exposure to high sunlight. These include changes in the size and composition of the pigment antennae, increases in the levels of the photosynthetic electron transport carriers, as well as an increase in the content of RuBisCO, the key enzyme involved in CO_2 fixation. Chloroplasts of leaves from plants that grow in high light environments contain greater quantities of carotenoids, ascorbate, and enzymes involved in ascorbate metabolism than leaves of shade plants. Adjustments to the pool size of the light-harvesting antenna protein, phosphorylation and spill-over of excess energy from PSII to PSI, are all mechanisms which help alleviate photoinhibition.

The xanthophyll cycle   One of the major photoprotective processes that facilitates the dissipation of excess energy within the antenna system is the

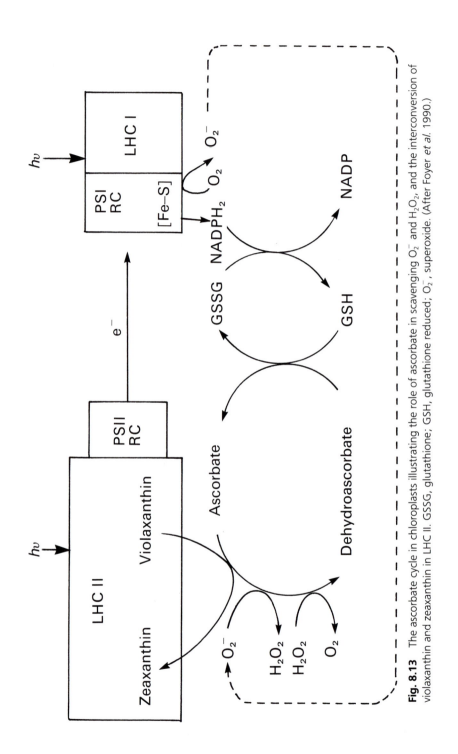

Fig. 8.13 The ascorbate cycle in chloroplasts illustrating the role of ascorbate in scavenging O_2^- and H_2O_2, and the interconversion of violaxanthin and zeaxanthin in LHC II. GSSG, glutathione; GSH, glutathione reduced; O_2^-, superoxide. (After Foyer *et al.* 1990.)

xanthophyll cycle. At very high PFD, the violaxanthin of the antenna is converted to zeaxanthin:

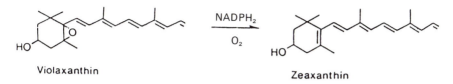

Violaxanthin

Zeaxanthin

This de-epoxidation requires an acidic thylakoid lumen pH and also ascorbate. Quantitative analysis of chlorophyll fluorescence quenching shows that as the level of irradiance of a leaf is raised, a compensatory relationship develops between photochemical quenching (q_P; a measure of non-cyclic election transport) and non-photochemical quenching (q_N; thermal quenching). One of the major components of q_N is energy-dependent quenching (q_E) that results from changes in the thylakoid membrane due to a pH gradient (ΔpH) – this q_E occurs mainly in the antenna chlorophyll and is dependent on zeaxanthin formation. The Mehler reaction, which supports linear electron flow from PSII to oxygen at PSI (§5.5), induces the development of a large ΔpH that in turn promotes zeaxanthin formation. The superoxide O_2^- and H_2O_2 generated during the electron flow are scavenged by ascorbate and reduced glutathione, so providing protection from the toxic oxygen radicals. Hence, in situations of excess irradiance, ascorbate plays a dual role: detoxification of products of the Mehler reaction, and secondly, stimulation of the xanthophyll cycle which in turn decreases the PSII activity. The ascorbate cycle is illustrated in Fig. 8.13. Thus, a combination of thermal quenching of excitation energy and photochemical quenching can mostly offset the effect of excess light. The factors which regulate these processes are pH, the zeaxanthin content, the phosphorylation state of LHC II, the activation state of ATP synthase, and the level of stromal metabolites including orthophosphate.

Sensitivity to photoinhibition in plants can be enhanced by other stress factors such as low or high temperatures, drought, high salinity, etc., which generally decrease the CO_2 fixation rates. Under field conditions, such stresses play an important role in plant metabolism. Many of the causes and effects of photoinhibition have been deduced from studies of short-term exposure of thylakoids and chloroplasts or even leaves to very high light. The data obtained from these experiments should be extrapolated with caution in predicting the behaviour of plant canopies which are exposed to varying light and dark periods under field growing conditions.

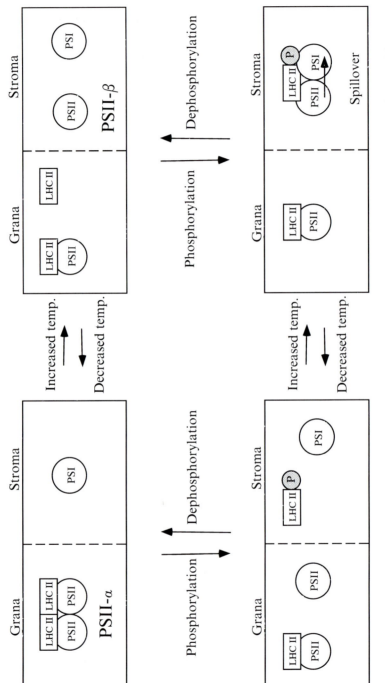

Fig. 8.14 The effects of LHC II phosphorylation and of temperature changes on the distribution of pigment–protein complexes and their functional interactions. (Courtesy: J. F. Allen, University of Lund.)

8.14   Energy redistribution between the two photosystems

The Z scheme of photosynthesis envisages that for photosynthesis to proceed efficiently there must be equal input of light energy to the reaction centres of both photosystems. This is not a problem at saturating light intensities. But at low light intensities or with light of a limited spectral range, an imbalance can occur and one of the photosystems could be preferentially excited over the other. However, O_2 evolution measurements and fluorescence data have shown that the steady state quantum yields for photosynthesis are constant and maximal over a wide spectral range where PSII absorption predominates. This suggests the existence of a mechanism in the photosynthetic apparatus for redistributing the energy absorbed by the light-harvesting pigments between the two photosystems so as to optimize the electron flow.

The process of temporary redistribution of excitation energy between PSII and PSI to ensure maximum quantum efficiency of photosynthesis is referred to as *state transitions*. When the two photosystems are almost equally excited and there is no *spillover* of energy from PSII to PSI, the system is said to be in state 1. When PSII is overexcited compared to PSI and there is transfer of part of the excitation energy to PSI, the system is said to operate in state 2. The phenomenon of state transitions has been observed in all O_2-evolving photosynthetic organisms. Plants growing in shaded areas, and algae and cyanobacteria living in ocean depths appear to have adapted these transition mechanisms very efficiently in order to maximize their photosynthetic capacity depending on the available spectral quality of the incident light.

How does the photosynthetic apparatus adapt itself to the variations in incident light intensities? It has been known for some time that spillover is accompanied by conformational changes in the chloroplast membrane. Biophysical (fluorescence kinetics and quantum yield), biochemical (^{32}P-labelling and fractionation of membrane proteins), and ultrastructural (freeze-fracture electron microscopy) studies initiated in the last decade have provided some insights into the mechanism of regulation of the state 1–state 2 transitions.

There is experimental evidence to suggest that in higher plants and algae the state 1 to state 2 transition involves the lateral (sideways) movement of some LHC II from the $PSII_a$ subunits in the grana stacks towards the PSI in the stromal thylakoids (Fig. 8.14). When PSII is overexcited, one or more of the 25–27-kDa polypeptides of LHC II becomes phosphorylated at a threonine residue by the transfer of a phosphate group from ATP to threonine. The reverse transition from state 2 to state 1 is associated with the dephosphorylation of LHC II. A specific membrane-bound kinase catalyzes the phosphorylation and a phosphatase catalyzes the dephosphorylation. Phosphorylation alters the net electrical charge of the LHC II located on the membrane surface, which presumably reduces the membrane–protein adhesion forces holding the grana stacks together. The phosphorylated LHC II detach themselves from the PSIIα subunits and move towards the non-appressed membrane domain in which the PSI is located (Fig. 8.14). Conditions which cause an increased demand for ATP (over and above the amount synthesized during non-cyclic photophosphorylation), for example by PSI photophosphorylation, may also induce LHC II phosphorylation and the state 1 to 2 transition.

Thermal excitation may also cause the movement of PSII particles from the grana stacks to the stroma region (PSIIα→PSIIβ). The phosphorylation of LHC II in isolated thylakoids causes a change in PSII Chl a fluorescence yield at room temperature as well as at 77 K, and hence can be monitored by measuring the decrease in the yield of PSII fluorescence emission at 685 and 695 nm relative to that of PSI fluorescence at 735 nm.

What is the molecular mechanism which triggers the light-dependent protein phosphorylation and which regulates the state transitions? It is reasonable to assume that the biochemical control mechanism should be located between PSII and PSI. Since, in high light, there is an excess generation of reduced plastoquinone in the quinone pool attached to the cyt bf complex, it was thought that LHC II phosphorylation is regulated by the redox state of the plastoquinones. According to this concept, the phosphokinase is activated by reduced PQ (PQH_2) and initiates LHC II phosphorylation. When the energy input to the two photosytems is balanced by a state transition, the PQH_2 is more rapidly oxidized to PQ (as a result of increased electron flow through PSI), the phosphokinase gets deactivated, and the non-phosphorylated LHC II is maintained in the grana stacks. This concept, however, does not explain how the phosphatase activity is induced after state transitions.

Recent evidence indicates that the cyt bf complex may be the regulator of LHC II phosphorylation. Mutants of *Lemna perpusilla* and of *Chlamydomonas reinhardtii* lacking the cyt bf complex have been isolated. The chloroplasts from these mutants possess normal membrane structure and have the same content of PSII and PSI reaction centres, LHC II and PQ pool as do their wild types. However, the cyt bf-less mutants do not undergo state transitions and their LHC IIs are not phosphorylated in light, although the phosphorylating kinase enzyme is present in the mutants. These observations pinpoint the cyt *bf* complex as the regulatory site for LHC II phosphorylation. The cyt *bf* complex is already known to play a vital role in optimizing the rate of photosynthesis by adjusting ATP production and NADP reduction through balancing the relative extent of non-cyclic electron flow (generating ATP and $NADPH_2$) and cyclic electron flow (ATP only) through the complex itself.

8.15   Role of light in the regulation of photosynthesis: the ferredoxin–thioredoxin control system

The activities of many enzymes of the CO_2 fixation pathway are modulated by light. These include NADP-glyceraldehyde-3-phosphate dehydrogenase, fructose 1,6-bisphosphatase, sedoheptulose 1,7-bisphosphatase, RuBisCO, phosphoribulokinase, pyruvate phosphate dikinase and malate dehydrogenase. Several mechanisms have been proposed to explain the light activation of these enzymes such as change in stromal pH, increase in stromal Mg^{2+} concentration, change in redox state, and the mediation of protein factors and 'effectors'. Only the reduced forms (with vicinal SH groups) of enzymes such as fructose bisphosphatase, sedoheptulose bisphosphate, and glyceraldehyde 3-phosphate dehydrogenase are active in the Calvin cycle. Their activity therefore is governed by the competition for electrons between PSI electron acceptors (NADP and O_2) and the enzymes themselves.

Illumination of chloroplasts increases the stromal pH from 7 to 8 or higher (due to uptake of protons from the stroma to the thylakoids) and increases Mg^{2+} concentration (due to a counter-movement of Mg^{2+} from the thylakoids to the stroma). Both increases enhance the catalytic activities of the above-mentioned enzymes of the Calvin pathway. The activation of RuBisCO is affected also by the chaperone rubisco activase (a light-modulated protein) and by other factors such as concentrations of sugar phosphates, CO_2, and reductants generated by photosynthetic electron transport.

Light modulation of the C_4 pathway enzyme pyruvate dikinase, which catalyzes the reaction:

pyruvate + Pi + ATP $\rightleftharpoons$ phosphoenol pyruvate + AMP + pyrophosphate

involves the interconversion between an active, non-phosphorylated form and an inactive, phosphorylated form of the enzyme. The level of ascorbate

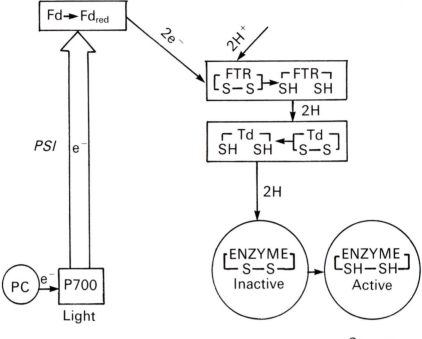

Fig. 8.15   Schematic representation of the role of the ferredoxin–thioredoxin system in the light-modulated regulation of chloroplast enzyme activities. FTR, ferredoxin–thioredoxin reductase; Td, thioredoxin. (After Buchanan in Barber *et al.*, 1992.)

in the stroma is high and the redox state of glutathione and its reductase may also play a role in the regulation of chloroplast metabolism.

The ferredoxin–thioredoxin system, comprising three proteins, ferredoxin (Fd), ferredoxin–thioredoxin reductase (FTR) and thioredoxin (Td), regulates the activities of at least four enzymes of the Calvin cycle (§6.2). Thioredoxins are low molecular weight proteins found in plants, animals and bacteria. Both thioredoxin and FTR, an iron–sulphur protein, contain reversibly reducible -S-S- linkages in their cystine residues. In the light, Fd is reduced at PSI, the reduced ferredoxin transfers electrons to FTR, and the reduced FTR in turn reduces thioredoxin (Fig. 8.15). Reduced thioredoxin activates the -S-S- containing enzymes FBPase, sedoheptulose bisphosphatase, phosphoribulokinase, and NADP-glyceraldehyde phosphate dehydrogenase of the CO_2 assimilation pathway. In the dark, all the enzymes are oxidized at their -SH groups and become deactivated. This regulatory effect of light on the enzymes ensures that CO_2 assimilation takes place in the day and carbohydrate degradation occurs mainly at night. Fd–Td system may also control chloroplast ATP synthase activity (§5.4).

8.16   Whole plant studies and bioproductivity

The efficiency of photosynthesis of the whole plant is crucial to agriculture, forestry, ecology, etc. when it comes to analyzing productivity for food and fuels and many other product uses. The quality and quantity of incident light (PAR), temperature and water stresses, availability and utilization of mineral nutrients, photorespiratory losses, presence of pollutants in the atmosphere (NO_2, SO_2, O_3) and in the soil (heavy metals), etc., are some of the factors which affect plant productivity. How these factors interact with the changing environment is now the subject of much practical and basic research.

The amount of incident radiation that can be captured by the plant will depend on its canopy and leaf structure. Leaf area is measured with a leaf-area meter. One commercial instrument (Plate VIII) employs the interruption of parallel light beams as leaves are passed between a light source and a sensor to measure the leaf surface area. Photosynthesis is usually saturated at moderate light intensities. Photosynthesis will function effectively, but not necessarily at its maximum capable rate, at a PPFD of 200 μmol m^{-2} s^{-1}, about one-tenth of the light encountered by plants when the sun is directly overhead on a cloudless day. In addition, many plants have the capacity to adapt to different light intensities during growth. Shade and sun leaves have the same photosynthetic efficiency (quantum yield measured under light-limiting conditions, usually 50–100 μmol m^{-2} s^{-1}). However, because shade leaves have less dark respiration, they achieve more net photosynthesis under low light conditions, i.e. they have a lower light compensation point than sun leaves. At high light intensities, sun leaves come into their own and show rates of photosynthesis many times those of shade leaves.

The difference in photosynthetic behaviour of shade and sun leaves is due to a number of factors. These include greater leaf thickness and internal surface area, larger mesophyll area, chlorophyll and more RuBisCO per unit area, and greater electron transport per unit chlorophyll. Information about the relative content of the two photosystems in leaves (number of antennae and reaction centre chlorophylls, grana and stroma lamellae, etc.) can be obtained by ultrastructure analysis and by absorption spectroscopy of the chloroplasts isolated from the leaves. These analyses show that the PSII:PSI ratios are higher in shade-adapted species which receive far-red-enriched ($\lambda \geq 700$ nm) light, than in sun-adapted species which receive far-red-depleted ($\lambda < 700$ nm) illumination. Thus the specific aspects of chloroplast structure, composition and function are determined by the light conditions available for plant growth.

Temperature is another important factor determining plant productivity – all photosynthetic reactions with the exception of the primary photoacts are thermochemical. Quantum yields are the same at 30°C for both C_3 (optimal temperature) and C_4 plants; for C_4 plants (growing in the tropics) the optimum is usually at a higher temperature. There is a complex relationship between temperature and net photosynthesis. Both dark and light respiration (which use up photosynthetic products) are also temperature dependent, but in a different manner to gross photosynthesis itself. Additionally, high and low temperature extremes, even for short periods, affect the subsequent rates of photosynthesis. Nevertheless, the normal physiological range for net photosynthesis is quite broad, with temperate and tropical species showing different optima.

Availability of fixed nitrogen is yet another factor contributing to plant productivity. Nitrogen is taken up from the soil as NH_4^+ or NO_3^-, and in the leguminous species by direct fixation of atmospheric N_2 by soil bacteria (Rhizobia) symbiotically associated with root nodules. The nitrogen status of the leaf directly influences its protein and therefore enzyme content and thereby affects the rate of photosynthesis. Nitrate which has been translocated from the roots to leaves can be reduced to ammonia by the photosynthetic apparatus under certain conditions, thereby providing a direct means of incorporating nitrogen into the leaf metabolism.

Many species of cyanobacteria also fix atmospheric N_2. One such N_2-fixing cyanobacterium which has application in agriculture is *Anabaena azollae*, which occurs in natural populations as a symbiont with an aquatic fern, *Azolla*. The *Azolla–Anabaena* symbiosis is widely used as a source of nitrogen fertilizer in rice fields throughout Asia. *A. azollae* can be separated from the host fern by mechanical means and then grown in the laboratory in the presence of synthetic foam pieces (polyvinyl or polyurethane). Under these conditions, the cyanobacterium adheres to the foam pores (Fig. 8.16) and is immobilized. Such foam-immobilized *A. azollae* maintain their photosynthetic activities for very long periods – N_2 fixation (ammonia release) being observed for several months. The spores of the cyanobacterium can stay dormant in the dry foam for years!

The importance of compiling productivity data from simultaneous measurements of all the integrated photosynthetic activities of whole leaves (or canopies), rather than individual reactions of cells or chloroplasts, is being recognized now. Instruments are available for studying photosynthe-

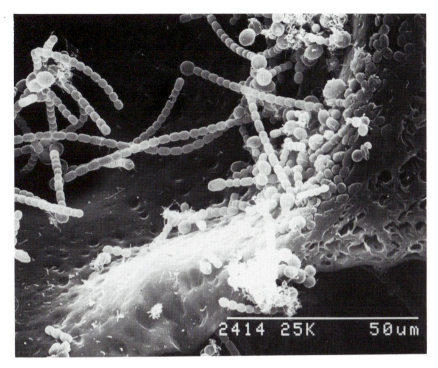

Fig. 8.16   Scanning electron micrograph of the cyanobacterium *Anabaena azollae* immobilized in polyvinyl foam pieces. (Courtesy: D. J. Shi and G. Morgan.)

sis in illuminated leaves with the simultaneous determination of O_2 and CO_2 exchanges and fluorescence emission (§1.8). Researchers use these techniques to study the effects of stresses such as excess light, high and low temperatures, water deficiency (drought stress), nutrient limitation, and the effect of salinity and added nutrients (e.g. phosphate and nitrate) on whole-leaf photosynthesis. The techniques are also useful for rapidly assaying the comparative photosynthetic activities of wild, hybrid and mutant strains of plants. Such non-destructive probes will be useful tools for crop physiologists and plant breeders in their work to optimize plant growth for certain soils and climates.

8.17   Photosynthesis and the greenhouse effect

A greenhouse is warmer than its surroundings because sunlight coming in through its glass roof and walls is absorbed by the plants and gases inside, raising the interior temperature. The *infra-red radiation* generated does not pass through the glass cover of the greenhouse and so is trapped inside the glass as heat. This phenomenon is known as the *greenhouse effect*. The planet earth is a big greenhouse in space, but instead of a glass roof it has built itself a blanket of water vapour and other 'greenhouse gases' to trap infra-red radiation as heat and keep itself warm.

The global annual use of fossil fuels (coal, oil and gas) is now equivalent to about 10 billion (10^9) tonnes of coal. The combustion of these fossil fuels and the burning of biomass (wood, straw, etc.) generate large quantities of particulate matter (fly ash), 'greenhouse' gases (oxides of carbon, sulphur and nitrogen, and including ozone and chlorofluorocarbons) and heat. Human activities now release into the troposphere an estimated 2.5 billion tonnes of particulate matter (fly ash and aerosol), 180 million tonnes of sulphur dioxide, and 40 million tonnes of oxides of nitrogen. This is in addition to the release into the atmosphere of CO_2 equivalent to about 6.0 billion tonnes of carbon from burning fossil fuels and 1.6 billion tonnes of carbon from land-use changes (mostly deforestation).

The CO_2 content of the atmosphere is increasing by approximately 1.8 ppm every year, and has gone up from 270 to 355 ppm (1992 figure) in the span of a century and is estimated to reach about 525 ppm in 50 years' time if we continue to generate CO_2 on past trends. Increasing concentrations of greenhouse gases are expected to cause a significant warming of the global climate in the next century – the effect being felt more at the high latitudes than in the tropics. Computer simulations of model ecosystems forecast temperature rises of 1.5 to 4.5°C for every doubling of CO_2 concentration. These temperature changes and accompanying variations in rainfall will occur unevenly over the global surface, affecting some countries and regions more than others. If these predicted climatic changes come true, then they would have a profound effect on global ecosystems and agriculture. Naturally this has created great concern among scientists, politicians and others, and has resulted in global negotiations to limit greenhouse gas emissions and to improve energy-use efficiency.

Is there any reason for alarm? Even though the indiscriminate release of particulate matter is detrimental to the environment, we should be aware of the potential beneficial effects of CO_2 especially, and NO_2 to a lesser extent, on photosynthetic productivity. Net photosynthesis by plants is promoted by higher levels of ambient CO_2. Increased CO_2 levels also result in better water and fertilizer use by plants and lower the photorespiration rate (in C_3 plants). Oxides of nitrogen are oxidized to nitrates and increase soil fertility. Short-term increases in leaf photosynthesis may or may not be translated into increased productivity of the whole plant at the end of the growth season. Changes in rainfall patterns will occur according to climate model predictions; thus water availabilities may alter both during the season and on an annual basis. As yet we understand very little about the complex interrelations which exist in plants and ecosystems when CO_2 and temperature levels are increased over the long term. More physiological, agricultural and ecological research is urgently required if we are to avert any possible negative consequences of the greenhouse effect. Remote sensing (using satellite) studies are now being undertaken to monitor biomass production in plant canopies (from measurements of scattered non-photosynthetically active radiation 700–1100 nm) and for the estimation of the predicted effects of CO_2 concentration on water availability. It is evident that a combination of climate and biological research of all types will be necessary if we are to understand how the future productivity of plants will be affected by climate change.

8.18   Mimicking photosynthesis

Photochemists and photobiologists are actively searching for synthetic systems which will split water using solar energy. The advantage of these artificial systems over natural photosynthesis may be that they might be optimized for maximum photosynthetic efficiency since they are not limited by the inherent physiological characteristics and requirements of whole plants. To be effective, such artificial devices mimicking biological processes (biomimetic systems) should be able to carry out the essential steps in natural photosynthesis, viz. light absorption and energy migration, charge separation, and electron transfer and catalysis. Ruthinium bipyridyl complexes and metalloporphyrins absorb light of visible wavelengths, have high extinction coefficients, have suitable redox potential differences between the ground and excited states, and good stability in light, and therefore are generally used as photosensitizers. Incorporation of the photosensitizer and electron donor in micelles or emulsions enhances the stability of the charged species generated after photon capture.

Light-induced electron transport has been demonstrated in synthetic membranes made up of liposomes embedded with pigments, proteins and other catalysts. Homogeneous or heterogeneous systems consisting of an electron donor, photosensitizer, electron mediator and hydrogen activation catalyst (platinum or hydrogenase) are now available for the photoproduction of hydrogen from water via a PSI-type of electron transport.

Photo-oxidation of water can be achieved using ultrafine semi-conductor particles of TiO_2 in the presence of RuO_2.

An optically transparent film of 15 nm TiO_2 particles sensitized with a monolayer of a charge transfer dye, trimeric ruthenium, was shown by Graetzel's group in Lausanne to be able to harvest 46% of the light flux and to convert $> 80\%$ of the incident protons to electric current. The overall light to electric energy conversion yield reported was 7–8% in simulated solar light and 12% in diffuse daylight.

A functional model of the O_2-evolution complex comprised of Mn porphyrin dimers was recently reported to catalyze O_2 evolution through four-electron oxidation of water.

One of the difficulties encountered in photoactive biomimetic systems is the high back-reaction rates, i.e. the rapid recombination of the photogenerated charged species before electron transfer to an acceptor can be accomplished. Some success towards overcoming this problem has been achieved by the synthesis of a tripartite molecule by Gust and co-workers (University of Arizona) which mimics the properties of a natural photosynthetic reaction centre. This so-called molecular triad consists of a porphyrin derivative (chromophore) sandwiched between a carotenoid (e^- donor) and a quinone (e^- acceptor) via covalent linkages. Laser flash excitation of this molecule generated within 100 picoseconds a charge–transfer complex (from the singlet state of the excited porphyrin) with a lifetime in the microsecond time scale – the back reaction was 100 times slower than the forward reaction. The quantum yield for the formation of the complex was about 25% and the energy stored in the complex was > 1 eV above the ground state.

Another active research field is the photoreduction of CO_2, to storable compounds, in the presence of dyes or semiconductors.

9

Laboratory experiments

9.1 Reference books for experiments

Hall, D. O., Scurlock, J. M. O., Bolhar-Nordenkampf, H. R., Leegood, R. C. and Long. S. P., eds. (1993). *Photosynthesis and Production in a Changing Environment: a Field and Laboratory Manual.* Chapman & Hall, London.

Hipkins, M. F. and Baker, N. R. (1986). *Photosynthesis Energy Transduction – a Practical Approach.* IRL Press, Oxford.

Packer, L. and Douce, R. (eds) (1987). *Methods in Enzymology*, Vol. 148: *Plant Cell Membranes.* Academic Press, New York.

Packer, L. and Glazer, A. N. (eds) (1988). *Methods in Enzymology*, Vol. 167: *Cyanobacteria.* Academic Press, New York.

Ross, C. W. (1974). *Plant Physiology Laboratory Manual*, 2nd edn. Wadsworth, Belmont, California.

San Pietro, A. (ed.) (1980). *Methods in Enzymology*, Vol. 69, Part C: *Photosynthesis and Nitrogen Fixation.* Academic Press, New York.

Walker, D. A. (1988). *The Use of the Oxygen Electrode and Fluorescent Probes in Simple Measurements of Photosynthesis*, 2nd edn. Oxygraphics Ltd, Sheffield, UK.

Witham, F. H., Blaydes, D. F. and Devlin, R. M. (1986). *Exercises in Plant Physiology.* Prindle, Weber and Schmidt, Boston, Mass.

9.2 Photosynthesis in whole plants and algae

(a) Measurements of photosynthetic CO_2 exchange using infra-red gas analyzer. Hall *et al.* (1993) Ch. 9; Witham *et al.* (1986) Exp. 41.

(b) $^{14}CO_2$ incorporation into leaves. Hall *et al.* (1993) Ch. 15; Ross (1974) Exp. 35.

(c) Measurement of O_2 evolution and chlorophyll fluorescence in leaves. Hall *et al.* (1993) Chs. 10 & 12; Walker (1988) pp. 17–46; Hipkins & Baker (1986) Ch. 4.

(d) Measurement of sugars and starch in leaves. Hall *et al.* (1993) Ch. 15.

(e) Measurements of activities of Calvin cycle and other enzymes involved in photosynthesis. Hall *et al.* (1993) Ch. 16; Witham *et al.* (1986) Exp. 40; Ross (1974) Exp. 36.

(f) Measurement of CO_2 compensation point in C_3 and C_4 plants. Ross (1974) Exp. 37.

(g) Light effect on rates of photosynthesis. Walker (1988) pp. 72 & 73; Witham *et al.* (1986) Exps. 37 & 38.

(h) Quantum yield of leaf photosynthesis. Hall *et al.* (1993) Ch. 10; Walker (1988) pp. 52–61.

(i) Batch culture techniques for algae and cyanobacteria. Hall *et al.* (1993) Ch. 21; Packer & Glazer (1988) pp. 68–95.

(j) Photoinhibition and O_2 evolution in algae. Hall *et al.* (1993) Ch. 21.

9.3   Preparation of protoplasts, chloroplasts and subchloroplast membranes

(a) Preparation of protoplasts from leaves. Hall *et al.* (1993) Ch. 7; Hipkins & Baker (1986) Ch. 2; San Pietro (1980) pp. 69–84; Walker (1988) pp. 136–140.

(b) Preparation of chloroplasts. Hall *et al.* (1993) Ch. 7; Hipkins & Baker (1986) Ch. 2; San Pietro (1980) pp. 85–104; Walker (1988) pp. 107–116; this book, §3.1 and Fig. 5.5.

(c) Isolation of mesophyll and bundle-sheath cells from leaves of C_3 and C_4 plants. Hall *et al.* (1993) Ch. 17; San Pietro (1980) pp. 55–68.

(d) Preparation of photosystem I and photosystem II subchloroplast fractions from spinach leaves. Hipkins & Baker (1986) Ch. 2; San Pietro (1980) pp. 129–141.

(f) PSI and PSII preparations from cyanobacteria. Packer & Glazer (1988) pp. 263–280.

(g) Reaction centre complexes from photosynthetic bacteria. San Pietro (1980) pp. 155–178.

9.4   Separation and estimation of photosynthetic pigments and proteins

(a) Separation of chloroplast pigments by solvent partitioning techniques. Ross (1974) Exp. 34; Witham *et al.* (1986) Exp. 28.

(b) Separation of pigments of photosynthetic membranes by high performance liquid chromatography (HPLC). Packer & Douce (1987) pp. 350–382.

(c) Estimation of chlorophyll. Hall *et al.* (1993) Ch. 16, Packer & Glazer (1988) pp. 274–322; Witham *et al.* (1986) Exp. 29; this book, §3.2.

(d) Electrophoretic separation and analysis of chloroplast membrane polypeptides. Hall *et al.* (1993) Ch. 18; Hipkins & Baker (1986), Ch. 3; San Pietro (1980) pp. 363–374; Packer & Glazer (1988) pp. 263–269.

9.5   Measurement of photosynthetic electron transport using oxygen electrode and/or spectrophotometer

(a) Whole-chain (PSII and PSI) electron transport from water to ferricyanide measured as O_2 evolution in the absence and presence of NH_4Cl: Estimation of 'intactness' of chloroplasts. Hall *et al.* (1993) Ch. 18; Hipkins & Baker (1986) Ch. 5; Walker (1988) pp. 122–125.

(b) Whole-chain electron transport from water to methyl viologen measured as O_2 uptake (Mehler reaction) in the presence of sodium azide. Hipkins & Baker (1986) Ch. 5; Walker (1988) pp. 126–128.

(c) Determination of the Hill reaction in chloroplasts by photoreduction of the dye dichlorophenolindophenol (DCPIP) measured spectrophotometrically. Ross (1974) Exp. 34, Witham *et al.* (1986) Exp. 39.

(d) Determination of PSII electron transport from H_2O to the dye p-phenylene diamine in the presence of ferricyanide measured as O_2 evolution. Hall *et al.* (1993) Ch. 18; Hipkins & Baker (1986) Ch. 5.

(e) Determination of PSI electron transport as O_2 uptake in DCMU-treated chloroplasts with ascorbate as electron donor. Hall *et al.* (1993) Ch. 18, Hipkins & Baker (1986) Ch. 5; Walker (1988) pp. 127 & 128.

(f) CO_2-dependent O_2 evolution. Hipkins & Baker (1986) Ch. 5; Walker (1988) p. 136.

9.6   Proton flux and photophosphorylation

(a) Measurement of transmembrane proton flux in broken chloroplasts using a pH electrode. Hall *et al.* (1993) Ch. 18; Hipkins & Baker (1986) Ch. 6.

(b) ATP formation from orthophosphate and ADP measured as the rate of alkalination of chloroplasts in light using a pH electrode. Hall *et al.* (1993) Ch. 18; Hipkins & Baker (1986) Ch. 6.

(c) Incorporation of ^{32}P-labelled orthophosphate into ATP by chloroplasts in light (photophosphorylation). Hipkins & Baker (1986) Ch. 6; San Pietro (1980) pp. 577–584.

Chemical names

The names of many of the chemical substances discussed in the text have undergone changes during the last few years; the following list may be useful.

Old name	New name
Acetaldehyde	Ethanal
Acetic acid	Ethanoic acid (old name still acceptable)
Acetylene	Ethyne
Citric acid	2-hydroxypropane-1,2,3-tricarboxylic acid
Ethyl alcohol	Ethanol
Ethylene	Ethene
Fumaric acid	*Trans*-butanedioic acid
Glutamic acid	2-aminopentanedioic acid
Iso-citric acid	1-hydroxypropane-1,2,3-tricarboxylic acid
α-ketoglutaric acid	1-oxobutanedioic acid
Malic acid	2-hydroxybutanedioic acid
Malonic acid	Propanedioic acid
Oxaloacetic acid	2-oxobutanedioic acid
Oxalosuccinic acid	1-oxopropane-1,2,3-tricarboxylic acid
Pyruvic acid	2-oxopropanoic acid
Succinic acid	Butanedioic acid

Abbreviations and prefixes used in the text

(a)	Abbreviations	Full term(s)
	Bchl	Bacteriochlorophyll
	BPheo	Bacteriopheophytin
	CAM	Crassulacean acid metabolism
	CF_0, CF_1	Chloroplast coupling factors
	Chl	Chlorophyll
	Cyt	Cytochrome
	DCMU	Dichlorophenyl dimethyl urea
	E	Einstein (one mole quanta of photons)
	EM	Electron microscopy
	EPR	Electron paramagnetic resonance
	Fd	Ferredoxin
	FNR	Ferredoxin–NADP reductase
	FTR	Ferredoxin–thioredoxin reductase
	IRGA	Infra-red gas analyzer
	LHC	Light-harvesting complex
	PAR	Photosynthetically active radiation
	PEP	Phosphoenol pyruvate
	Pheo	Pheophytin
	PPFD	Photosynthetic photon fluence density
	PQ	Plastoquinone
	PS	Photosystem
	Q	Quantum yield
	q	Quenching of fluorescence
	q_P	Photochemical fluorescence quenching
	q_N	Non-photochemical fluorescence quenching
	q_E	Energy-dependent fluorescence quenching

RC	Reaction centre
RuBisCO	Ribulose bisphosphate carboxylase
psa	Genes coding PSI proteins
psb	Genes coding PSII proteins

(b) *Unit prefix*

Unit prefix	*Meaning of the term*
giga (g)	10^9
kilo (k)	10^3
micro (μ)	10^{-6}
milli (m)	10^{-3}
nano (n)	10^{-9}
pico (p)	10^{-12}
femto (f)	10^{-15}

Further reading

Non-specialist books

Jones, H. G. (1992). *Plants and Microclimate*, 2nd edn. Cambridge University Press, Cambridge.

Lange, D. O., Nobel, P. S., Osmond, C. B. and Ziegler, H. (eds.) (1981–83). *Physiological Plant Ecology, I–IV*. Encyclopedia of Plant Physiology, Vols. 12A to 12D. Springer-Verlag, Berlin.

Lawlor, D. W. (1993). *Photosynthesis: Metabolism, Control and Physiology*, 2nd edn. Longman, Harlow, England.

Nobel, P. S. (1991). *Physicochemical and Environmental Plant Physiology*. Academic Press, San Diego.

Raven, P. H., Evert, R. F. and Eichhorn, S. E. (1992). *Biology of Plants*, 5th edn. Worth Publishers, New York.

Stanier, R. Y., Ingraham, J. L., Wheelis, M. L. and Painter, P. R. (1987). *General Microbiology*, Macmillan Education, London.

Walker, D. A. (1992). *Energy, Plants and Man*, 2nd edn. Oxygraphics, Brighton, UK.

More specialized books

Baker, N. R. and Bowyer, J. R. (eds.) (1994). *Photoinhibition of Photosynthesis: From molecular mechanisms to the field*. Bios Scientific Publishers, Oxford.

Barber, J. (ed.) (1976–1992). *Topics in Photosynthesis*, Vols. 1–11. Elsevier, Amsterdam.

Barber, J., Guerrero, M. G. and Medrano, H. (eds.) (1992). *Trends in Photosynthesis Research*. Intercept Ltd, Andover, UK.

Beadle, C. R., Long, S. P. Imbamba, S. K., Hall, D. O. and Olembo, R. J. (1985). *Photosynthesis in Relation to Plant Production in Terrestrial Environment.* Tycooly Publications, Oxford.

Coombs, J., Hall, D. O. and Chartier, P. (eds.) (1983). *Plants as Solar Collectors.* D. Reidel Publishing Company, Dordrecht.

Govindjee, Amesz, J. and Fork, K. (eds.) (1986). *Light Emission by Plants and Bacteria.* Academic Press, New York.

Hall, D. O., Scurlock, J. M. O., Bolhar-Nordenkampf, H., Leegood, R. C. and Long, S. P. (eds.) (1992). *Photosynthesis and Production in a Changing Environment: A Field and Laboratory Manual.* Chapman & Hall, London.

Houghton, J. T., Callender, B. A. and Vourney, S. K. (eds.) (1992). *Climate Changes 1992.* Cambridge University Press, Cambridge.

Kirk, J. T. O. (1994). *Light and Photosynthesis in Aquatic Systems.* 2nd edn. Cambridge University Press.

Murata, N. (ed.) (1992). *Research in Photosynthesis,* Vols. 1–4. Kluwer Academic Publishers, Dordrecht.

Packer, L. and Glazer, A. N. (eds.) (1988). *Methods in Enzymology,* Vol. 167: *Cyanobacteria.* Academic Press, London.

Pearcy, R. W., Ehleringer, J. R., Mooney, H. A. and Rundel, P. W. (1989). *Plant Physiological Ecology: Field Methods and Instrumentation.* Chapman & Hall, London.

Staehelin, L. S. and Arntzen, C. J. (eds.) (1986). *Photosynthesis III. Encyclopedia of Plant Physiology,* Vol. 19. Spring-Verlag, Berlin.

Stolz, J. F. (ed.) (1991). *Structure of Phototropic Prokaryotes.* CRC Press, Boca Raton.

WRI (1992). *World Resources: a Guide to the Global Environment 1992/93.* World Resources Institute/Oxford University Press, Oxford.

Reviews and articles

Annual Review of Plant Pathology and Plant Molecular Biology

Bennet, J. (1991). Protein phosphorylation in green plant chloroplasts. **42**, 281–311.

Bowes, G. (1993). Facing the inevitable: Plants and increasing atmospheric CO_2. **44**, 309–332.

Bowler, C., Van Montague, M. and Inze, D. (1992). SOD and stress tolerance. **43**, 83–116.

Brown, R. H. and Bouton, J. H. (1993). Physiology and genetics of interspectific hybrids between photosynthetic types. **44**, 435–456.

Demmig-Adams, B. and Adams III, W. W. (1992). Photoprotection and other responses of plants to high light stress. **43**, 599–626.

Flügge, U. I. and Heldt, H. W. (1991). Metabolite translocators of the chloroplast envelope. **42**, 129–144.

Furuya, M. (1993). Phytochromes: their molecular species, gene families, and functions. **44**, 617–645.

Glazer, A. N. and Melis, A. (1987). Photochemical reaction centres: structure, organization and function. **38**, 11–45.

Golbeck, J. H. (1992). Structure and function of photosystem I. **43**, 293–324.

Krausse, G. H. (1991). Chlorophyll fluorescence and photosynthesis: the basics. **42**, 313–349.

Long, S. P., Humphries, S. and Falkowski, P. G. (1994). Photoinhibition of photosynthesis in nature. **45**, 633–662.

Lumsden, P. J. (1991). Circadian rhythms and phytochrome. **42**, 351–371.

Ort, D. R. and Oxborough, K. (1992). In situ regulation of chloroplast coupling factor activity. **43**, 269–291.

Potrykus, I. (1991). Gene transfer to plants: assessment of published approaches and results. **42**, 205–225.

Somerville, C. R. (1986). Analysis of photosynthesis with mutants of higher plants and algae. **37**, 467–507.

Spreitzer, R. J. (1993). Genetic dissection of RuBisCO structure and function **44**, 411–434.

Stitt, M. (1990). Fructose 2–6 bisphosphate as a regulatory molecule in plants. **41**, 153–185.

Thompson, W. F. and White, M. J. (1991). Physiological and molecular studies of light-regulated genes in higher plants. **42**, 21–53.

Vermaas, W. (1993). Molecular-biological approaches to analyze photosystem II. **44**, 457–482.

Woodrow, I. E. and Berry, J. A. (1988). Enzymatic regulation of photosynthetic CO_2 fixation in C_3 plants. **39**, 533–594.

Biochimica Biophysica Acta

Allen, J. F. (1992). Protein phosphorylation in regulation of photosynthesis. **1098**, 275–335.

Aro, E., Virgin, I. and Andersson, B. (1993). Photoinhibition of photosystem II. Inactivation, protein damage and turnover, **1143**, 113–134.

Debus, R. J. (1992). The manganese and calcium ions of photosynthetic oxygen evolution. **1102**, 269–352.

Demmig-Adams, B. (1990). Carotenoids and photoprotection in plants: a role for the xanthophyll zeaxanthin. **1020**, 1–24.

Genty, B. Briantais, J. M. and Baker, N. (1989). The relationship between the quantum yield of photosynthetic electron transport and quenching of chlorophyll fluorescence. **990**, 87–92.

Gutteridge, S. (1990). Limitations of the primary events of CO_2 fixation in photosynthetic organisms: the structure and mechanism of rubisco. **1015**, 1–14.

Melis, A. (1990). Dynamics of photosynthetic membrane composition and function. **1058**, 87–106.
Van Grondelle, R., Dekker, J. P., Gillbro, T. and Sundstrom, V. (1994). Energy transfer and trapping in photosynthesis. **1187**, 1–65.

New Phytologist

Furbank, R. T. and Foyer, C. H. (1988). C_4 plants as valuable model experimental systems for the study of photosynthesis. **109**, 265–277.
Nobel, P. S. (1991). Achievable productivities of certain CAM plants: basis for high values compared with C_3 and C_4 plants. **119**, 183–205.
Walker, D. A. (1992). Excited leaves. **121**, 325–345.
Woodward, F. I. (1992). Predicting plant responses to global environmental change. **122**, 239–251.

Photochemistry and Photobiology

Bolton, J. R. and Hall, D. O. (1991). The maximum efficiency of photosynthesis. **53**, 545–548.
Friesner, R. A. and Won, Y. (1989). Photochemical charge separation in photosynthetic reaction centres. **50**, 831–839.
Lagoutte, B. and Mathis, P. (1989). The photosystem I reaction centre: structure and photochemistry. **49**, 833–844.
Vogelman, T. C. (1989). Penetration of light into plants. **50**, 895–902.
Zuber, H. (1985). Structure and function of light-harvesting complexes and their polypeptides. **42**, 821–844.

Photosynthesis Research

Anderson, J. M. (1992). Cytochrome b_6f complex: dynamic molecular organization, function and acclimation. **34**, 341–357.
Blankenship, R. E. (1992). Origin and early evolution of photosynthesis. **33**, 91–111.
Buchanan, B. B. (1992). Carbon dioxide assimilation in oxygenic and anoxygenic photosynthesis. **33**, 147–162.
Calvin, M. (1989). Forty years of photosynthesis and related activities. **21**, 3–16.
Coleman, W. J. (1990). Chloride binding proteins: mechanistic implications for the oxygen-evolving complex of photosystem II. **23**, 1–27.
Duysens, L. N. M. (1989). The discovery of the two photosystems: a personal account. **23**, 131–162.

Foyer, C. H., Furbank, R., Harbinson, J. and Horton, P. (1990). The mechanisms contributing to photosynthetic control of electron transport by carbon assimilation in leaves. **25**, 83–100.

Giersch, C. and Krausse, G. H. (1991). A simple model relating photoinhibitory fluorescence quenching in chloroplasts to a population of altered photosystem II reaction centres. **30**, 115–121.

Gray, J. C. (1992). Cytochrome f: structure, function and biosynthesis. **34**, 359–374.

Hansson, O. and Wydrzynski, T. (1990). Current perceptions of photosystem II. **23**, 131–162.

Hatch, M. D. (1992). I can't believe my luck. **33**, 1–14.

Horton, P. and Ruban, A. V. (1992). Regulation of photosystem II. **34**, 375–385.

Nickell, L. G. (1993). A tribute to Hugo P. Kortschak: The man, the scientist and the discoverer of C_4 photosynthesis. **35**, 201–204.

Oquist, G. (1992). On the relationship between the quantum yield of photosystem II electron transport, as determined by chlorophyll fluorescence and the quantum yield of CO_2-dependent O_2 evolution. **33**, 51–62.

Vermaas, W. F. J. (1994). Evolution of heliobacteria: Implications for photosynthetic reaction center complexes. **41**, 285–294.

Witt, H. T. (1991). Functional mechanism of water splitting. **29**, 55–77.

Physiologia Plantarum

Allen, J. F. (1995). Thylakoid protein phosphorylation state 1–state 2 transitions, and photosystem stoichiometry adjustment: redox control at multiple levels of gene expression. **93**, 196–205.

Asada, K. (1992). Ascorbate peroxidase – a hydrogen peroxide scavenging enzyme in plants. **85**, 235–241.

Baker, N. R. (1991). A possible role for photosystem II in environmental perturbations of photosynthesis. **81**, 563–570.

Foyer, C. H., Lelandais, M. and Kunert, K. J. (1994). Photooxidative stress in plants. **92**, 696–717.

Krall, J. P. and Edwards, G. E. (1992). Relationship between photosystem II and CO_2 fixation in leaves. **86**, 180–187.

Stitt, M. and Quick, P. W. (1989). Photosynthetic carbon partitioning: its regulation and possibilities for manipulation. **77**, 633–641.

Scientific American

Bazzaz, F. A. and Fajer, E. D. (1992). Plant life in a CO_2 rich world. **266**, (1), 18–24.

Gasser, C. S. and Frawley, R. T. (1992). Transgenic crops. **266**, (6), 34–39.

Govindjee and Coleman, W. J. (1990). How plants make oxygen. **262**, (2), 42–51.

Youvan, D. C. and Marrs, B. (1987). Molecular mechanisms of photosynthesis. **256** (6), 42–48.

Trends in Biochemical Sciences

Arnon, D. I. (1987). Photosynthetic CO_2 assimilation by chloroplasts: assertion, refutation, discovery. **12** 39–42.

Arnon, D. I. (1988). The discovery of ferredoxin: the photosynthetic path. **13**, 30–33.

Barber, J. and Andersson, B. (1992). Too much of a good thing: light can be bad for photosynthesis. **17**, 61–66.

Deisenhoffer, J., Michel, H. and Huber, R. (1985). The structural basis of photosynthetic light reactions in bacteria. **10**, 243–248.

Knaff, D. B. (1988). Reaction centres of photosynthetic bacteria. **13**, 157–158.

Knaff, D. B. (1991). Regulatory phosphorylation of chloroplast antenna proteins. **16**, 82–83.

Lavergne, J. and Joliot, P. (1991). Restricted diffusion in photosynthetic membranes. **16**, 129–134.

Nitschke, W. and Rutherford, A. W. (1991). Photosynthetic reaction centres: variations on a common structural theme. **16**, 241–245.

Rochaix, J. D. and Erickson, J. (1988). Function and assembly of photosystem II: genetic and molecular analysis. **13**, 56–59.

Rutherford, A. W. (1989). Photosystem II, the water-splitting enzyme. **14**, 227–232.

More specialized articles

Andersson, I. Knight, S., Schneider, G., Lindqvist, Y., Lindqvist, T., Branden, C. I. and Lorimer, G. H. (1989). Crystal structure of the active site of ribulose-bisphosphate carboxylase. *Nature*, **337**, 229–234.

Barber, J. and Anderson, B. (1994). Revealing the blueprint of photosynthesis. *Nature*, 370, 31–34.

Deisenhoffer, J. and Michel, H. (1989). The photosynthetic reaction centre from the purple bacterium *Rhodopseudomonas viridis*. Nobel Lecture. *EMBO Journal*, **8**, 2149–2170.

Hartman, F. C. and Harpel, M. R. (1994). Structure, function, regulation and assembly of D-ribulase-1,5-bisphosphate carboxylase/oxygenase. *Annual Review of Biochemistry*, **63**, 197–234.

Ikeuchi, M. (1992). Subunit proteins of photosystem I. *Plant Cell Physiology*, **33** (6), 669–676.

Scurlock, J. M. O. and Hall, D. O. (1991). The carbon cycle. *New Scientist*, **51**, 1–4.

Seibert, M., DeWit, M. and Staehelin, L. A. (1987). Structural localization of the O_2-evolving apparatus to multimeric (tetrameric) particles on the lumenal surface of freeze-etched photosynthetic membranes. *Journal of Cell Biology*, **104**, 2257–2265.

Index

A_0 49, 82, 166, 167
A_1 49, 82, 166, 167
absorption spectra 58, 64
  of carotenoids 44
  of chlorophyll 41
  of photosynthetic bacteria 129
  of phycobiliproteins 45
  and resonance transfer 64
accessory pigments 40, 68, 84
action spectrum 65, 68
adenosine diphosphate (ADP) 30, 87–9, 91,
    132, 198
adenosine triphosphate (ATP) 29, 30, 79–83,
    87–9, 91–5, 99, 102, 107–10, 113, 132,
    134, 198
adenosine triphosphatase (ATPase) 74, 88, 150
Agrobacterium 152, 153
algae 1, 100, 106, 123, 140, 149
  blue-green 43
  brown 40
  green 40, 99, 123, 150
  red 40, 43, 45, 138
allophycocyanin 40, 45, 46
Anabaena azollae 151, 190–91
antenna pigments 51, 52, 163
Arabidopsis 152–3, 173
atrazine 97–8
autoradiography 101, 180
Azolla 190

bacterial photosynthesis 126
bacterial reaction centres 128, 131–3, 138–42
bacteriochlorophylls 127 130, 132, 136–8
bacteriopheophytin 132–3
bacteriorhodopsin 135
BChl *see* bacteriochlorophylls

biomass 5, 192–3
biomimetic systems 194
bioproductivity 189
BPheo *see* bacteriopheophytin
bromoxynil 97
bundle sheath cells 54–6, 111, 113, 115

C_3 plants 56, 124–5
  general characteristics 116–17, 121–2
C_4 plants 56, 123–5
  CO_2 fixation by 111–14
  general characteristics 116–17
  photosynthetic apparatus of 53–6
Ca ions (in O_2 evolution) 78, 162–3
Calvin cycle 29, 99–106
CAM (crassulacean acid metabolism) 114–17
carbon dioxide
  compensation point 117, 125
  cycle 3
  in greenhouse effect 192–3
carbon dioxide fixation 99, 111
  by C_3 pathway 99–106
  by C_4 pathway 111–14
  by CAM 114–17
  by photosynthetic bacteria 133–5
  in dark and light 104
    energetics of 107–08
  experimental techniques 195–6
carotenoids 43, 44, 130
CCCP 89
Ceramium (red alga) 37
CF_0 88, 89
CF_1 88, 89, Plate I
chaperones 155
chemiosmotic hypothesis (Mitchell's) 30, 86,
    87

Chlamydomonas 85, 150, 174
Chlorella 25, 27, 65, 99, 102
chloride ions (in O_2 evolution) 78, 162, 163
Chlorobiaceae 126, 128
Chlorobium 126, 128, 130, 131
Chloroflexaceae 126, 128
Chloroflexus 126, 128, 130, 131, 142, 143
chlorophyll *a*/*b* protein complex 74, 154
chlorophylls 41
    absorption spectra of 41
    determination of 42, 43
    fluorescence of 60–62, 158
    formulae of 41
    structure of 42
chloroplasts 32
    electron transport in 82, 197
    envelope 38, 153–4
    genetics 150–51
    isolation of 37–9
    origin and development of 148–9
    pigments 39
    structure 33–7, 54–6, 156–7
    transport of polypeptides 153–6
    types of 37
chlorosomes 130–31
Chromatiaceae 126, 128
Chromatium 126, 128–9
chromatophores 127
chromosome of rice chloroplast genome 151
Coccomyxa (green alga) 37
coupling factor (see CF_0 and CF_1)
cyanobacteria 40, 136–7, 148–9, 162
cyanophora (cyanelle) 148
cyclic photophosphorylation 81, 91
cytochrome 49, 84
    b_{559} 49
    b_6 49, 70, 82, 170
    f 49, 70, 82, 84, 170
    b/*f* complex 49, 82, 170–72

DCMU 71, 86, 96, 97, 176, 197
D_1–D_2 proteins 74, 76, 139, 140, 180
difference spectrophotometer 71
donor–acceptor species 62, 63

Einstein 7, 9
electromagnetic radiation 5, 6, 9
electron acceptors and donors 72, 95, 96
electron microscope (EM) 15–21
electron paramagnetic resonance (EPR) 15, 16
electron spins 13, 57
electron transport
    bacterial 132–3
    comparison of plant and bacterial 137–41
    in chloroplasts 79, 81, 82, 91, 197
electron volt 8
electroporation 153
Emerson 30

effect 67, 68
flashing light experiments 27, 52
energy levels of electrons 57
energy transfer 64
    between photosystems 184–6
Engelmann's experiment 23, 24
EPR spectra 16, 160, 168
etioplasts 149
evolution of photosynthesis 142–3
excited states: singlet, triplet 57–60

F_A, F_B and F_X 49, 81, 82, 167
fatty acid biosynthesis 119
ferredoxin (Fd) 16, 50, 81, 82, 92, 93, 118, 120, 132, Plate I
ferredoxin-NADP reductase (FNR) 49, 82, 85, 165, Plate I
fluorescence 13, 58–60, 175, 179, 196
    delayed 63
    induction (Kautsky) curves 175–6
    lifetime 61
    low temperature (77 K) 61, 62, 158, 177
    maximum 175–78
    of *Spirulina* Plate V
    variable 175–8
fluorometer 14, Plate VII
freeze-fracture EM 19, Plate I
fructose bisphosphates 109–11, 187

glucose 107, 110
glutathione 119, 182–3
glycollate (glycollic acid) 106, 121–2
grana 32–4, 48, 156, 158
greenhouse 191
    effect 191–3
    gases 192

halobacteria 135
heat shock proteins (Hsp) 155
Heliobacteria 126–8, 130, 131
Hill reaction 28, 197
Hill and Bendall (Z) scheme 70, 83
hydrogen peroxide 92, 94, 120–21, 182

infra red gas analyzer (IRGA) 11–13, 191, 195, Plate VI
inhibitors 95
    of electron transport 86, 96, 97
    of energy transfer 91
    of oxygen evolution 96
    of photophosphorylation 89, 91
ionophores 91
ion transport in chloroplasts 155
iron sulphur centres 49, 167

Kautsky effect 175
Kortschak, Hatch–Slack pathway 31, 111
Kranz-type anatomy 53, 116

lamellae 32–5
lateral migration 185
light absorption and emission 58, 59
light and dark reactions 26, 27, 30, 79
light and phytochrome action 145–7
light as radiation; units and terms used 9
light-coupled reactions other than CO_2
    assimilation 118–21
light-harvesting pigment complexes 44
light regulation of photosynthesis 187–8
lipids 51
lumen
    in measurement of light 8
    in thylakoid structure 33, 170–72
lux 8

malate (malic acid) 112–15
manganese in oxygen evolution 78, 159–63
Mehler reaction 92, 197
membrane complexes 82, Plate I
membrane electrical potential 87
mesophyll cells 54–6, 111–15
mimicking photosynthesis 193
Mitchell 30, 86–7

NADH$_2$ 132, 134
NADPH$_2$ 29, 30, 79–84, 86, 87, 91–5, 102,
    103, 105, 107, 108, 156
nigericin 91
nitrate reductase 118
nitrite reductase 118
nitrogenase 118
nitrogen assimilation 118
nitrogen fixation 118, 190
non-cyclic electron transport 81–3
Nostoc flagelliforme (cyanobacteria) 20

oxaloacetate (oxaloacetic acid) 112–15
oxygen
    cycle 3
    electrode 10, 11, 90, 197
    evolution 65–7, 76–8, 159, 161–2, 196–7
    exchange 10, 11, 197
    measurement 10, 11, 197
oxygen-evolving complex 48, 76–8, 159, 162

P680 48, 52, 76, 77, 81, 82, 159, 163, 164, 180
P700 49, 52, 72, 81, 82, 166
P840 128, 132
P870 128, 129, 132
P890 128, 129
PAGE (polyacrylamide gel electrophoresis)
    74, 75, 180
PAR (photosynthetically active 400 to 700 nm
    radiation) 4, 9, 124
PEA 14, Plate VII
PEP (phosphoenol pyruvate) 112, 113, 115
PEP carboxylase 112, 113, 115
phosphoglyceric acid (PGA) 100, 102–05

phosphinothricin 97
phosphorescence 59, 61
photoinhibition 179–83
photon 7–9
photorespiration 121–3
photosynthesis
    definition 1
    effect of CO_2 25, 26, 124–5
    effect of light intensity 25, 26, 66, 124–5
    effect of temperature 25, 26, 125
    history of 22–31
    mimicking 193
    whole plant 189
photosynthetic
    apparatus 32–5, 53–6, 127
    bacteria 126
    efficiency 67
    electron transport and phosphorylation 79
        in bacteria 132
        cyclic 91, 92, 170
        non-cyclic 81–3
    photon flux density *see* PPFD
    reaction centres 52, 72, 74, 86, 128, 129,
        131, 138, 142, 163
    unit 51
        of bacteria 130, 132
        of chloroplasts 51, 52
photosystems I and II 68–70, 72–4, 82, 83,
    156, 159, 162, 163, 165–7
phycobilins 40, 43
phycobilisome 46, 44, 138
phycocyanin 40, 45, 46
phycoerythrin 40, 46
phylloquinone 46, 167
phytochromes 145–7
plastocyanin (PC) 50, 82, 165
plastoquinone (PQ) 47, 48, 51, 82
pmf (proton motive force) 87
polypeptides
    chloroplast 75, Plate I
    envelope 154
    light-harvesting complex 163, 166
    oxygen-evolving complex 163
    of PSI and PSII 163, 166, 167
    synthesis and transport 153–6
Porphyridium 70, 71
PPFD 9
primary electron acceptors 49, 72, 133, 166
protein phosphorylation 184–6
proton translocation 170, 172
protoplasts 148, 153
PSI and PSII *see* photosystems I and II
PSII$_a$ and PSII$_\beta$ 156, 157
psa and *psb* gene products 151, 167, Plate I

Q cycle 170–72
quanta 7
quantum efficiency (yield) 65–7, 108, 178
quantum radiometer 10

quantum requirement 65, 66, 108
quantum theory (Planck's) 6
quenching of fluorescence 177–8

redox potentials (E_m) 48, 69, 80, 81
reduction and oxidation 80
reductive carboxylic acid cycle 134
Rhodobacter sphaeroides 128–9, 132
Rhodopseudomonas viridis 128–9, 132–3
Rhodospirillaceae 126, 128
Rhodospirillum rubrum 128, 129
ribulose bisphosphate (RuBP) 69, 81, 82
ribulose bisphosphate carboxylase (RuBisCO)
        103, 121–3, 171, 173–4
Rieske (Fe–S_R) protein 49, 82, 171–2
RuBisCO X-ray structure of Plate IV
rubisco activase 173

S states in water oxidation 63, 77, 159, 161
scanning electron microscopy (SEM) 17, 18
sensitized fluorescence 64
separation of
    chloroplast pigments 197
    photosystems 72–4
solar energy, efficiency of capture and
        turnover 4, 5
spillover of energy between photosystems 185
Spirogyra 23, 24
starch synthesis 109–11
State 2–State 1 transitions 102
Stoke's shift 63, 64
stomata 115, 117
stroma 30, 32, 35, 172
sucrose synthesis 109, 112
sulphate assimilation 118, 119

superoxide 120, 182, 183
superoxide dismutase 120

T-DNA 152
thermoluminescence 63
thioredoxin (Td) 89, 188
thylakoids 32–5, 73
transgenic plants 152
    generation of 153
    by *Agrobacterium* vector 153
    by biolistics 153
    by electroporation 153
    by microinjection 153
    by microlaser 153
translocator (C_3 phosphate) 109, 156
transmission electron microscopy 17–20
triose phosphate 102–05, 110–12

ubiquinone 47, 130, 133, 139
uncouplers 89

valinomycin 89
vesicles 75–6
violaxanthin 182, 183

water oxidation (*see* oxygen evolution)

xanthophyll 43, 44
xanthophyll cycle 181, 183

Y_z 48, 76, 77, 163, 165

'Z' scheme 30, 70
zeaxanthin 182, 183